FROM THE WORLD'S LEADING SCIENTIST BEHIND
'HOW TO DEVELOP A TOP 10 TENNIS RANKING'

MY [TENNIS]

FEDERER. NADAL.

BLACK BOOK

Vol. I

SERENA. DJOKOVIC.

NAVIGATE & PLAN TOWARDS A TOP 10 RANKING

DR. ASHLEY M. B

AM8 International

Beyond Top 10 Tennis

Beyond Top 10 Tennis Rankings

Your name is to complete the 'blank' that follows **Federer, Nadal, Serena, Djokovic**, __________. **My [Tennis] Black Book** has been built especially for *you* to follow in their footsteps, and it starts from the beginning. Be sure to write your name atop each of these pages. After all, this is __________ Black Book. Let's begin!

AM8International.com

FROM THE WORLD'S LEADING SCIENTIST BEHIND
'HOW TO DEVELOP A TOP 10 TENNIS RANKING'

MY [TENNIS]

FEDERER. NADAL.

BLACK BOOK

VOL. I

SERENA. DJOKOVIC.

NAVIGATE & PLAN TOWARDS A TOP 10 RANKING

DR. ASHLEY M. BERGE, PHD

First published in Australia in 2024 by AM8 International

https://AM8International.com

AM8 International
Gold Coast, Australia

https://AM8International.com

1st Edition
ISBN: 978-0-6450597-4-8

It's kind of fun to do the impossible.
– Walt Disney

Malibu & Maldive, my superheroes to guardian angels.
As always, this wouldn't have been possible without you by my side, one word at a time.

Foreword

My [Tennis] Black Book has been a long time in the making. As with all my previous books, this is from me to you and to help you progress towards the pinnacle of play. In reality, it is so much more. Before penning **My [Tennis] Black Book** I had to ask myself some serious questions. Most of these questions revolved around if I was in your shoes – what I would *want* or what would I more importantly *need*, unbeknownst to me, to help me on my Pathway towards that Top 10 tennis ranking.

First off, let's rewind 12 months ago to the release of 'How to Develop a Top 10 Tennis Ranking – The Power of the 8th Key' and how this very book was the culmination of exhaustive and extensive research. Little did the reader truly know that it was in fact the final 'extract' from my doctoral thesis I had completed some 7 years prior. Of importance is that whilst this was the closing of one chapter, in a manner of speaking – molding a doctoral thesis into a readable text with real-world application and understanding for players, coaches and/or parents alike, over the past 8 years since this milestone I've continued to build upon this body of work – relentlessly, and this 'new' research has continued to feature in each and every book I've penned to date. **My [Tennis] Black Book** is, however, the very first addition in what marks a new chapter and milestone itself.

Serving as a middle ground between the former and the new due to the absolute depth of knowledge and application shared across 11 *non-fiction* Books, **My [Tennis] Black Book** has been designed with these works at the forefront whilst also forming its own cornerstone that allows you to progress towards that *next* Level of play. No one said progressing towards a Top 10 tennis ranking let alone achieving Grand Slam success would be easy. But neither did they think it would be possible to follow a Pathway that would nudge them closer every step of the way. What I've come to realise, nonetheless, is that whilst this Pathway exists over the course of these 11 Books – 'I am Your Tennis Coaching Guru' provided the very first Pathway of its kind from a coaches perspective – from me to you, it was in fact penned for the coach to follow and to guide their player over the course of *all* Books that had come prior. At that stage, there were 8 Books and

'I am Your Tennis Coaching Guru' formed No. 9 – the first guidepost to unify these works and help put them into action.

What wasn't available then is available now. Courtesy of 'The 7 Keys to Optimise Your Life' and 'How to Develop a Top 10 Tennis Ranking' in conjunction with the Beyond Top 10 Tennis Podcast, presents a huge step forward to truly unlock new milestones and integrate these into your game with the ultimate reward. As such, only now with *all* these Books finally out in the open has it become feasible to share the players guidepost – a light-post that is entirely *yours*. **My [Tennis] Black Book** is designed to be personal – just for you. It embodies those initial questions of what I would *want* and/or *need* if I was in your shoes and only those answers I continued to circle back to made the final cut. Built to be tangible, **My [Tennis] Black Book** does not discriminate. But it does ask for the knowledge of *all* previous Books – over time. However, if you do not have *all* this knowledge yet, do not worry – this has been integrated across *stages* and Levels.

As was done with the coach's guidepost in 'I am Your Tennis Coaching Guru – a 10 Year Pathway of Lessons and Planners for Junior to Amateur Players towards a Top 10 Ranking', the time has come for your very own – for the player, with a significant amount of 'recall' from all prior Books that comprise of the *Collection*. How has this become possible? Quite simply, courtesy of the power of science *fused* with my own personal playing experience and ultimately, my extensive coaching background – *plus* twenty years in the process, you're in good hands.

As a player, I once dreamed of achieving not just *the* Grand Slam but the Golden Slam to follow in the footsteps of Steffi Graf. Not even the likes of **Djokovic**, **Serena**, **Nadal** or **Federer** (at time of writing) – in order of their number of Grand Slam titles, were able to achieve this feat. And whilst **Djokovic** is the sole remaining active player and this season is **Nadal's** final season, both **Serena** and **Federer** have since called time. Of course, **Serena** (2012) did achieve the career Golden Slam as did **Nadal** (2008) with **Federer** coming close with a Silver (2012) and **Djokovic** a Bronze (2008) – *until* this year's 2024 Olympics. Courtesy of **Djokovic's** triumph and securing perhaps the biggest *win* that has since eluded him, you can now add **Djokovic** to the list of career Golden Slams. Most profoundly, **Djokovic** is the most senior of all three of these players to achieve this feat with both **Serena** and **Nadal** claiming their Olympic Gold in arguably peak periods of their career. Of course, it is noteworthy to state that

Djokovic has endured a lengthier peak period of play than *any* of these four players with 20+ Grand Slams to their name and rather his Gold has come towards the later end of his career in contrast to the earlier peak periods of both **Serena** and **Nadal's** careers, respectively. Interestingly and worthy of inclusion, is Murray who is the only player to achieve back-to-back Olympic Gold let alone *win two*, defeating **Federer** in the 2012 Olympic Gold medal match. And up until the 2024 Olympics, all four players with 20 or more Grand Slam titles had only one Olympic Medal to their name – **Djokovic's** Bronze, **Serena's** Gold, **Nadal's** Gold and **Federer's** Silver, with **Djokovic's** Gold this season making him the only one of these players to have more than one *singles* medal. Of course, **Serena**, **Nadal** and **Federer** have all won Olympic Gold in doubles which is something that has eluded **Djokovic** whilst **Serena** is the only one of these players to *win* four Gold medals – one singles and three doubles with nonother than her sister, Venus.

The *two* players who have achieved four consecutive Grand Slam *wins* although not considered *the* Grand Slam as they were *not* inside the same calendar year, are both **Serena** (2014-2015) and **Djokovic** (2015-2016). Neither **Nadal** nor **Federer** held all four Grand Slams at any one time. However, **Federer** made the final of all four Grand Slams during the 2006 and 2007 seasons with the French Open eluding him, whilst **Djokovic** achieved the same feat during the 2015, 2021 and 2023 seasons with the Australian Open being the only Grand Slam to *win* on *all* these occasions. It is Steffi Graf, however, that remains the sole player to achieve the true Golden Slam – something that these four players who remain the greatest of *all* time in their own right, were unable to achieve and thus trail Graf's *Golden* accomplishment. Needless to say, as a player I had wild dreams that I dreamt of turning into a reality.

As a coach, these dreams didn't alter course. Rather, they compounded. When I transitioned from player to coach, I had to ask myself what my core driver was and what ultimately, I'd aspire to achieve as a coach. Whilst there is various versions of what in varied forms, primarily these are along the lines of having one of my players *win* each Grand Slam inside a calendar year to achieve *the* Grand Slam as a coach. Another version is to achieve identical results but for my players to face off against one another in the final of *each* Grand Slam. And of course, for an Olympic year to have achieved the Grand Slam feat and for one of my players to *win* the elusive Gold to secure *the* coach Grand Slam. Pretty wild, right? To make it all even, let's add a Silver and Bronze there, too, for good measure because why not.

How would a coach be able to achieve this if only one player in the history of our sport has been able to do so? Well, I had to think out of the box and here we are. But before I get into more of the how, as has become custom in each and every Book that I have penned, here's a behind the scenes breakdown of *how I came to be:*

The how, *recap*:

At 18, I became a Personal Trainer / Sports & Strength Conditioning Coach. I was also a full-time athlete, training for 8 hours a day.

At 19, I was a sponsored Elite Athlete. I also commenced University and maintained my full-time training.

At 20, I became a Level 1 Coach, then went on to become one of Australia's youngest ever Level 2 Coaches permitted to go back-to-back.

Just before I turned 21, I received a scholarship to play and train at one of America's leading Academies, in Texas.

At 21, I was coaching some of America's top national players as a direct result of succumbing to another injury and whilst also playing and training when healthy. That same year I received a full scholarship to a National Collegiate Athletics Association (NCAA) DII private University that marked against the Harvard Grading Scale.

At 22, I spent part of the summer touring through Europe with American national players. At this time, I had also started my Undergraduate degree and had received a Grade Point Average (GPA) of a 4.0 (the highest attainable) in both first and second semesters. This came with multiple Dean's awards.

At 23, I again spent part of the summer touring through Europe with American national players. Due to injury, I had to fly back to Australia and leave my scholarship behind. I spent the majority of the year on crutches unable to walk. This was the year I completed my Undergraduate degree and started to run. This was also the year I had to choose between playing and coaching — I chose coaching.

At 24, I commenced my Masters. I spent the summer, again, touring through Europe with American national players. On return I was asked to be the Director of a tennis program and accepted, being one of the youngest coaches, nationally, to take on such a position. After running for less than a year I managed to clock up 50 km days (not weeks). Just before I turned 25, I managed my biggest week yet: 180 km's in total.

At 25, I registered my first business with my Dad and opted not to travel but to keep on top of my Masters (after receiving a GPA of 7.0: the highest attainable). Shortly after, my Dad passed away – just shy of my graduation; four months later, my Grandpa passed away.

At 26, I competed in my first Ultramarathon (100 km) but had a DNF (Did Not Finish) due to an ongoing hamstring injury. In the same year, I became recognised as a Level 3 Coach (the highest attainable, globally). Soon after, I commenced my PhD – I still had questions to answer that my studies until then were unable to address.

At 27, my PhD was under full-swing – in not one, but between *two* Schools (a result of the depth of the work undertaken) and I was on track to complete my PhD in the minimum timeframe. The same year I was able to sponsor my first athlete.

A fortnight after turning 28, I became unwell. Ultimately, I couldn't run, walk, read, write, do simple tasks, let alone finish the homestretch of my PhD without near-collapsing. Without going into detail, somehow, I did manage to finish my PhD and in the minimum timeframe; with only four months leave taken. For an elaboration, *The Secrets to Optimal Wellbeing* shares more.

At 29, I had PhD corrections to look after, although the entire year was all about getting my health back. That same year, I competed in my first Ultramarathon since becoming unwell – and I finished all 80 km's of it, even though I tore a tendon in my ankle at the 20-25 km mark. It was about finishing, and I did. Mid-year, I had surgery that could have left me far worse off (from my ongoing health battle) and I am incredibly fortunate to still be here.

At 30, I had already published the first instalment in *The Secrets to Optimal Success* series: *The Secrets to Optimal Performance Success*, and later that year, *The Secrets to Optimal Wellbeing* was published. And, almost two years since submission, I was officially awarded my doctorate. It was also a year of tightening the vision for what the science inside my PhD was, and eventually over time, going to bring the world. As a result, *The Science of Elite Performance* was published just shy of turning 31.

At 31, it was all about bringing this vision into fruition – and so the journey began of building a Social Compilation, Topicthread, that brought the world together, using the *two* disciplines of my doctorate, Health *and* Education, with *performance* its overarching tone, as the vehicle to unify individuals across the world. And with that, 'social' was at its core – using the scientific model inside my PhD. This saw Topicthread (topicthread.com) come to life.

At 32, *The Secrets to Optimal Coaching Success* was published, and slowly but surely, Topicthread was the World's 1st Social Compilation – and the 1st interface with the architecture (at the time) to deliver access to Health *and* Education, in unison, universally. Meanwhile, quietly in the background I had been working on *The Elite Research Method*, which was subsequently published later that year.

At 33, the first instalment of *What is Your Game Missing?* was released, although what followed was a bittersweet year. The final chapter was called on a three-year relationship, whilst I received a DNF, seven years after my first, in the pursuit of another 100km Ultramarathon I had trained *all* year to be a pacer for – calling it a day at the 50km mark with the onset of injury (a tear on the inside of my knee – MCL). As a result, delivering the second instalment of the series: *What is Your Game Missing, Now*? later the same year, was perhaps my most mentally exhaustive book to date with a mixture of emotions endured throughout. Despite these temporary hurdles since overcome, it was *this* very book that landed the Number One spot on the top Tennis Books to read in 2020 – a milestone in itself.

At 34, with *big* shoes to fill, the Education release embedded inside Topicthread coincided with the publication of an *eighth* book – third instalment of the same series: *What is Your Game Missing, to Win?* And as the year will go down in history as the beginning of a pandemic that slowly reshaped our world, an unanticipated *ninth* book was released: *I am Your Tennis Coaching Guru* to bring light and hope to those in need of a training Pathway over the *next* 10 Years of Play.

At 35, the year was immersed in developing a heightened Level of applied awareness, a practice that's been honed over the past eighteen months to afford a *new* depth of understanding to interwork with *new* scientific discoveries. The year couldn't have been more rewarding as a *tenth* book was released – *The 7 Keys to Optimise Your Life* marking a new Level of complexity endured, spanning the length of the entire year between varied shapes to concepts to ideals before the 7 Keys became apparent as overdue – to be shared. This served as a timeless reminder of how our behaviours shape our outcomes and that an optimal state is translucent when our behaviours synchronise to feed our performances and as an individual, we become the controllers of our own optimal states – with the 7 Keys lighting the way *towards* a Top 10 tennis ranking.

At 36, it was time to dare and to take a new risk. This resulted in my very own fictional imprint being born: Pink Octopus Books that went on to release my first fictional book the same year that embodied the lessons over the course of the past two years – where the book bore its beginnings. With a new age coming forth and with Women front and centre, we are beginning to have more of us at the head of the table and so it was only fitting that *The Power of Another Woman* was celebrated in its entirety. I'll definitely be back for more. On a professional note, I became Chairperson of the Board for a Not for Profit in the Mental Health space to play my part in spreading greater awareness towards our health in all its forms. And on a personal note, it was two-fold. Starting to run again after being sidelined for over a year due to plantar fasciitis (inhibiting the ability to walk, let alone run), brought me great joy, whilst I closed a private chapter in my life that resulted in much needed healing. This also coincided with two bouts of COVID. It goes without saying that The Power of Another Woman remains as so much more than a heart etched onto its pages — rather an embodiment of both hardship to broken hearts and the joy to epitome of love that awaits when we learn to set ourselves free.

At 37, I was drawn back to the body of work where the 8th Key had been kept in the shadows all along – with the time beckoning to set it free. Contrary to belief, this was a really hard decision. It was letting go of a piece of me that I had held dear. As a player and coach this would be the final piece of the puzzle – what was missing above all else, to truly achieve that Top 10 Tennis Ranking. And so, after much internal debate I decided that keeping it under lock and key wasn't helping anyone, not even me, and that the enormity of the work also deserved to be celebrated. 'How to Develop a Top 10 Tennis Ranking' paid tribute to what has come before it and the staggered-by-design approach – the time arriving for feathers to be ruffled and for voices kept quiet opting for no more. This brought to fruition not only the rallied cries of those who wished they knew, but Beyond Top 10 Tennis launched – a Podcast set on speaking the truth and relenting against the quiet to share additional insights interwoven behind the now 8 Keys as the vocal compendium to guide the '*best in the world*' towards the top echelon of play. Behind the scenes, my running came to a halt as I was diagnosed with hip impingement, most probably developed during my younger playing years, which temporarily sidelined me once again. But if there's anything you've learned from this *recap*, relentless I was and am to finish the year with a few strides under my belt to build on into the next year.

And now at 38, it has been a whirlwind of a rollercoaster. Beyond Top 10 Tennis has over 130 episodes. Our transcripts continue to amass with well over 1,000,000 words. It has been phenomenal and a heck of a lot of work whilst this year has also included a redesign of AM8 International to ensure all of this is accessible right at your fingertips. Beyond Top 10 Tennis has knuckled down on serving as a platform to discuss not only this body of work, but key implications with current WTA and ATP Tour Rankings & Results, allowing players to better understand where they may be on the Pathway and what Keys they may need to begin to and/or more firmly integrate into their game. Essentially, Beyond Top 10 Tennis has been pivotal in bringing this work further to life and when I asked myself what's next or what's missing – what's that *next* Book that is calling my name, it was this one. But do not be fooled, the initial half of the year was all about asking these questions relentlessly to figure out what was next – what was still missing, and how Beyond Top 10 Tennis could also help, too. As such, **My [Tennis] Black Book** was no accident. It's full of intent. And for those with a suspicious undercurrent or in favour of the 13th it is perhaps simple happenstance, or not, that **My [Tennis] Black Book** marks my 13th Book overall. But not without a price as it has been enduring in more ways than one. Whilst I was back running this year, still working towards where I was some years prior, ganglions were identified bilaterally in my wrists and penning No. 13 has been anything other than *easy*. But if I can be relentless in building out *your* Black Book and the entire framework to reach the top of the game, for you to follow, then you can be relentless in your pursuit, too. But none of this would be possible without my two little sidekicks – Malibu & Maldive.

Now that you've had access to a little more context, it is that much simpler to frame the earlier *how*. I had to do the *work*. If I was truly serious about setting the tennis world alight and to push boundaries that weren't even thought fathomable, I *had* to do the work. I had to become the world's leader. I had to make that transition from player to coach – which I'd already done, and then to academia. And I fought relentlessly, literally, for this to happen and to make it a reality.

Little do people know how close I was to not being able to commence my doctorate. I'd followed the rules and I'd done the work. But after I'd finished my Masters and went to commence my PhD I was stopped. That's right – they tried to stop me. They tried to say that my research component for acceptance was not met, however, the error was theirs – not factoring in my studies in the United States – on a *harder* scale. Fast-forward a year – an entire 12 months after being prevented entry, I wasn't about to give up. I went and gathered the support I needed, internally, with the right people that had my back. And it worked. Despite being granted acceptance within certain confines i.e. a probationary period as they anticipated failure, I went ahead and proved them wrong. I surpassed each milestone they set for me. Oh, and did I mention I completed my doctorate in the minimum timeframe? That's right. I pushed like no other because I knew how important the work I was doing really was and that I needed to get back out into the 'real world' and transition this 'work' into readable Books opposed to being confined to academic jargon, to ensure that players, coaches and/or parents would have access to these incredible findings.

No, I didn't think it would take an additional 10 Years. But when I had to break it all down, make the work accessible for *all*, this is a complete culmination of what it really takes to develop not just the best tennis players in the world, but to achieve milestones most only dream of. Now trackback to that initial *how* and the plausibility even a few minutes ago until now. Now you know the truth behind it and the scale – how it really has been made a reality and *how* I've fought to make this possible every step of the way – for *you*.

With **My [Tennis] Black Book** marking my 13th Book – 12 of these now guide you every step of the way towards the pinnacle of play. *Nine* years in the making and *twelve* since I initially started my doctorate plus the additional *two* thereafter of the aforementioned corrections to appease those who tried to prevent my milestone achievement – with my health compromised, these *nine* have been spent relentlessly writing to share this work with you whilst ensuring it remained legible, readable and understandable in contrast to being locked away in an academic journal and/or database for purely academics to access. As for that *one* in between, I was completely out of action due to my aforementioned health.

The irony here is that for the last 9 years I have battled with my University to keep my work under embargo until it was completed. If it was released prior, it may very well have been misused and yes there have been attempts over the years. It would also potentially be misinterpreted and do more harm than good. At the same time, it wouldn't allow me the freedom to write – to continue to research and uncover more *wins* from this data as I also embarked on numerous other adjoining studies that essentially overlapped the overarching premise of *how*.

The initial work laid the fundamental foundations. The most significant of these was a key indicator that directly correlated with a Top 10 tennis ranking. Even more so, implications for the Top 5 tennis players on the WTA and ATP Tours. But imagine putting a doctoral dissertation out into the world with those kinds of ramifications without being re-written into a digestible context. By that, I am referring to the language we use to speak to one another daily in contrast to what you may read in a scientific journal. However, this is far from the truth of the work that I've shared. There's a somewhat middle ground that needed to be built. A compromise of sorts. Why? Simply because a learning curve needed to be integrated into the work to ensure each one of you had the capacity in some way, shape or form to *learn* what was before you. By design, this has been cyclical as it feeds back into the data and *how* individuals learn.

From the very beginning this work followed suit, it was directed by the data and by design the language progressed from more rudimentary terms to the more complex scientific explanations. This further underpins the stepping stone analogy from one Book to the next and not by accident. Rather, an intentional design I needed to consider with each new release. And whilst there is quite a lot more to the backend of this approach, it has intermittently been shared in an explanatory nature in the two cornerstone releases that encapsulate the 8 Keys – 'How to Develop a Top 10 Tennis Ranking: the Power of the 8th Key' and 'The 7 Keys to Optimise Your Life: Using Tennis to Develop Behaviours that Deliver Optimal Performance in Play & in Life'. What's most fundamental here is that these Keys did *not* coexist with the initial doctoral dissertation. These Keys have been investigated and thoroughly put to the test over the past decade to break down the pivotal intersection between one key and the next, whilst collectively – again by design per the data, refining all Books down to *eight* distinct subsets that have colloquially been termed *Keys*. And if you've read 'How to Develop a Top 10 Tennis Ranking' then you too understand the power of language used and its refinement.

If I hadn't fought tirelessly over the past 9 years to keep this work under embargo, then these Keys would not have been made available nor would the additional *wins* that have continued to come out of this work. For example, the predictive analytics behind Grand Slam Champions would *not* be accessible nor the rulers behind Replicated Success – *how* to gauge if a player will go on to *win* more than one Grand Slam. The same applies for the Barrier Breaker to the Golden Rule and more (terms widely discussed in the *Collection*). Ultimately, all of these 'moving pieces' that have since been shared are the first of their kind and absolute in their premise – prerequisites to achieve a Top 10 tennis ranking, to *win* a Grand Slam to multiple, plus provide direct correlations with Olympic success.

That *how* is now really within reach. From that child that dreamed of the Golden Slam before injuries ran their course, to that Coach that dreamed of the 'Coaches' Golden Slam for her players, all this work has been a culmination to this point in time. The *how* is now available and so is that additional voice of reason. Together, **My [Tennis] Black Book** pulls on these tools to help you make your dreams a reality and in doing so contributes to bringing the once unfathomable to life.

But there's something else. This year marks 10 years since my health was hit for six. For those not familiar with that colloquial expression, a common straight forward surgery I underwent a decade prior due to an accident experienced as a child, went wrong. It would be a drastic understatement to say it flipped my life upside down, but it essentially has taken the best part of those 10 years to get a semblance of my life back, albeit quite different. Like so many who experience complications with their health, adjustments need to be made. This meant I had to rethink *how* I could help you when I no longer had the same physical capacity to be with you on court each step of the way.

The biggest irony here is that if I didn't need to think outside the box about the *how* that perhaps this body of work would not exist. But as I remained relentless in figuring out this how, it became clear that I could have an even greater impact by making my work accessible to every one of you in contrast to one player or one Club and/or Academy. Fast-forward to 2024 – a decade later, my collective health is the best it has been since that detrimental surgery. Whilst I am not as physically active as prior, perhaps running 150km's plus per week wasn't as sustainable as I liked to think and yet I am still running – albeit a third of that mileage currently and have more time for Malibu and Maldive who were not in my life all those years ago. But the last 6 and 4 years, respectively, has been a very different story as they've both patiently laid close by as I've penned all of the last now 9 Books and that's not something as easily attainable when the need to travel across the globe from one tournament to the next is the life of a player and their respective coach on either the WTA or ATP Tours.

Of course, I may be persuaded *now* for the right player, my intention remains concrete. If a player has followed the 8 Keys and thoroughly read the Complete *Collection*, then the answer is yes – my availability becomes more tangible in-person for you at Grand Slams. If the answer is no then my intention stands – do the initial work before bringing me in. This reaffirms that if I've put the work in then I expect the same from you, too. And if that's the case and Grand Slam *wins* set your heart alight, I am here to help.

My [Tennis] Black Book is now yours and built to harness the *secrets* dispersed throughout each Book that has paved the way until now. However, I'll be the first to admit that there is a *lot* of reading and a lot of work if you're just getting started. That's where **My [Tennis] Black Book** comes into play. Whilst there is no room for shortcuts nor no replacement for hard work, there is a more streamlined approach to cross check and to reference that will ensure you're either on track and/or remain on track towards your intended endpoint — be in securing a professional ranking on the WTA or ATP Tours, a local tournament *win*, a Final 8 result at a Grand Slam or to continue to add Grand Slams to your current tally. **My [Tennis] Black Book** is designed for you and to keep you accountable. Importantly, it is crafted exclusively from me to you with the underlying (or overarching) *how* front and centre.

If Grand Slams to a Golden Slam is in your now and/or your future, let's begin by making some waves and welcome to the next generation of play where the stage – or centre court, sooner or later, will become your playground. All I ask is one thing – that you put in the work and reach out when the time arrives to ensure that I can help push you that little bit further and bring those Grand Slam dreams into reality. And of course, if the Golden Slam is what you're also after, then times ticking — let's get started!

Contents

Tables

My [Tennis] Black Book

Federer, Nadal, Serena, Djokovic, __________

Navigate & Plan towards a Top 10 Ranking

Vol. I

Dr. Ashley M. Berge, PhD

My [Tennis] Black Book: Federer, Nadal, Serena, Djokovic, __________

My [Tennis] Black Book belongs to:

Version:

1 2 3 4 5

I am ____ years of age

My dream is to:

My favourite Grand Slam is:

I would like to Win _____ Grand Slam(s)

My biggest weapon is my:

This year, I'd like to:

In the next 10 Years, I'd like to:

Welcome

My [Tennis] Black Book is unlike any book I've penned before. There are similarities that coexist between the pages that can be identified in the *eleven* preceding Books – the expansive *Collection* to this point in time, but none have been as clear nor have provided the 'navigation' to earnestly consume Book 1 – '*The Secrets to Optimal Performance Success*' through to Book 11 – '*How to Develop a Top 10 Tennis Ranking*', whilst providing you with the space to reflect, recall and truly review at both a micro (weekly) and macro (seasonal) Level. But it has been far from easy. Encompassed within **My [Tennis] Black Book** is a multifaceted compendium that affords a means to *navigate* Books 1 through 11 whilst itself forms an immense resource of integrated readings dispersed throughout an entire *season*. And there's more. But what's unique to this release is the interwoven 'plans' across the same timespan you will be prompted *daily* to complete.

You will notice the scale and scope of the integrated planning. Whilst every effort has been taken to refine the details to avoid a sense of overwhelming whilst affording a sense of completeness, equally has the effort been taken to ensure **My [Tennis] Black Book** can be easily taken with you. This has resulted in a refined scale in contrast to a *larger* version to perhaps more effortlessly complete. In turn, a separate *Journal* has been made to provide you with ample room to complete all tasks. But to pack as much as you can muster into your Black Book, each Block provides room to ensure the balance between scale and scope remains intact in *addition* to the adjoining *Journal*. This you'll soon come to know and understand.

A lot of time has gone into contemplating *how* big is *too* big to *how* big is just the *right* fit. **My [Tennis] Black Book** accounts for all these subtle questions which are shared to signal the immense thought that has gone into the design process before you. I remain acutely aware that *if* the design of **My [Tennis] Black Book** was *too* 'this' or *too* 'that' then the likelihood of your 'rate of success' would become compromised. As such, it has been *you* who has remained at the forefront of these decisions to ensure **My [Tennis] Black Book** serves as your compendium to success.

Whether your goal is to *win a* Grand Slam or more, secure a Top 10 tennis ranking and/or to play on the *biggest* stages the tennis world has to offer, it all comes down to *you* and your Level of commitment. The choice is always yours. *Navigating* towards the pinnacle of play is now here courtesy of the unrelenting research and compiled data on the *best* in the tennis world across more than *twelve* years – *twenty* years collectively. This has been a *first* of its kind for tennis historically and now for you as the next generation embarks on their 'Pathway' towards a Top 10 ranking. You are the first generation to have access to this comprehensive framework to *navigate* and *plan* towards a Top 10 ranking. For you, it is different than generations prior. This time you've got data on your side. And this time, you have an embedded Pathway to help guide you and in plain print – Books 1 through 11 *plus* **My [Tennis] Black Book** to guide you along the way.

Learning is *key*. **My [Tennis] Black Book** is built for you to progressively *learn* how to progress towards the top of the game. It is also designed to nudge you to put in the work *each* Block and stay the course over the length of an entire season. If you're ready to embark on that Top 10 ranking, then **My [Tennis] Black Book** has your back. And if the *Golden* Slam lights a fire at your core, then once the *Collection* has made its way in your knowledge bank, you know where to turn.

Irrespective of your current ranking and/or Level of play, **My [Tennis] Black Book** has been built to accommodate all levels of the game with a common denominator – unequivocal commitment towards achieving a Top 10 ranking. Whether this season or next decade is not of concern, but rather if you stay the course your time will arrive. After all, it will be what you do *with* **My [Tennis] Black Book** that will differentiate one player to the next – eligible to join the 'Top 8%' this season or inside the next two will be up to *you*.

Of course, time will vary depending on your current 'State of Play', but the principles remain. When you consider the Top 100 players on the WTA and/or ATP Tour, that represents the current 100% of *elite* players and only 10% of those players represent the Top 10 of which 8% (or *less*) will maintain their exclusive hold for the remainder of the season, as established and shared in detail in 'How to Develop a Top 10 Tennis Ranking'. But when this range is extended to include rankings up to the Top 1000 – the Top 100 is then representative of the 10% and the Top 10 becomes the 1% and to maintain your ranking inside this exclusive range narrows to 0.8%. It is remarkable to consider that courtesy of science and the *twenty* years behind this work and extensive scientific modelling that *finally* this Pathway is yours and at your disposal to embark on the next generation of play *with* science on your side.

If you're currently sitting inside the Top 1000 rankings the first natural point of call will be the initial 10% before the 1% becomes within reach. Similarly, if you're currently inside the Top 100 then the Top 8% is your primary marker. Depending on your scope of view, both 0.8% and 8% remain relative across the rankings, however, these markers are representative of your 'rate of success' now and current trajectory. Scientifically, the Top 100 has been representative of *the* 100% and in return the Top 10 as *the* 10% but these are mere datapoints on your timeline and if your ranking is closer to the Top 1000 in contrast to the Top 10 then each milestone is relative to *you*. Your current 10% may be another players 1% but the destination is one and the same and that's what remains front and centre throughout **My [Tennis] Black Book**. This is *your* time and *your* journey towards the top of the game and each one of you will differ – slightly to significantly, on where you fall on that timeline. And that's how it is meant to be.

Only 2% of players – per the Top 100 limit for inclusion, will become a 'barrier breaker' each season and enter the Top 10 whilst the 8% will remain. If your scale is broader, then this equates to 0.2% and 0.8% respectively. Simply put, not all of you will enter the Top 10 inside the same season. That's the data speaking *and* the rankings. But per your time and your rate of success will serve as a primary indicator for *when* you enter these respective ranges which are ultimately hinged on your ability to accommodate and therefore integrate, all 8 Keys into your game. **Federer** had the 8 Keys. Then **Nadal** followed with *all* Keys before **Djokovic** harnessed the 8 Keys and delivered a new Level of frequency in application. All *three* of these players arrived at different times. At the same time, however, was **Serena** on the adjacent Tour with *all* 8 Keys and new benchmarks set along the way – leading the way for *both* Tours until **Djokovic** surpassed her Grand Slam tally. All *four* of these players have been pivotal in the development of the *Collection* and *your* Black Book as their results and rankings, respectively, have scientifically led the way, amongst other players to meet the '10%' rulers, in building the ultimate framework that delivers Top 10 rankings and more. This is where **My [Tennis] Black Book** serves its biggest lesson and hands the reins over to *you* to follow the footsteps of **Federer**, **Nadal**, **Serena** and **Djokovic** and ask yourself if your name will be *next*? To find out, it is *time* to commence.

How to Use Your Black Book

This is *your* Black Book to complete as you choose and only *you* can complete it – no one else. The structure of *your* Black Book is relatively streamlined from Week 1 through to Week 48 – spanning an entire season with four weeks *removed* to use as you choose. This does *not* mean you will be playing and/or training for the entirely of this period. It does mean you have the option to complete these weeks at your discretion. You may choose to *read* multiple weeks ahead of time. You may choose to *listen* ahead of time and/or you may choose to devour each Book of the *Collection* in advance.

What's important to remember is that **My [Tennis] Black Book** is akin to your sidekick. It is equally a means to *navigate* the 8 Keys that the *Collection* embodies. Whilst **My [Tennis] Black Book** is powerful as a standalone and rich in additional and invaluable content – it becomes your *superpower* when used in conjunction with the *Collection.* As such, *your* Black Book has been penned as this superpower and pulls on language to varied theories and/or terminology that you have already become acquainted with and have grown to understand from your prior *learnings* courtesy of the *Collection*. However, *fear not* for those of you who are using **My [Tennis] Black Book** as your first point of call. If there are words and/or phrases you're unfamiliar with, simply highlight these and/or make note and these terms will become known as you make your way through *your* Black Book.

It is important to remember as you make your way through *your* Black Book that there are multiple *weekly* requirements and some *daily* tasks to complete. It is recommended to dedicate at least *six* to *seven* hours as a ***minimum*** each week to complete *your* Black Book. Whilst this time will vary for each of you, this is a simple threshold set to ensure you get the most out of *your* Black Book. How you choose to use this time and when is not necessarily of concern but rather the *completeness* of your *weekly* requirements with one exception. Your *daily* tasks are encouraged to be completed directly after or as close as possible to the completion of your match-play (tournament or otherwise) and/or training session and to later be elaborated on in the adjoining *Journal*.

Whilst there are no set rules for *your* Black Book the layout has been designed to point you in the intended direction. It is up to you whether you choose to follow these directions or not. Due to the multifaceted design of **My [Tennis] Black Book**, there are various themes of inclusion – both **Chapters** and **Topics** to *read*, **Episodes** to *listen* to, Questions to *answer* and assorted Notes to be *taken*. *How* you choose to complete *your* Black Book from Week 1 through to Week 48 is your choice to make. Due to the volume of information available – directly inside your *Black Book* or through the corresponding *Collection*, it is by no means suggested that this will be covered in its entirety inside the length of a season. Rather, it is quite the contrast.

This is where your 'Point of Inception' (PoI) comes into play. If you're inside your *first* decade of play then it is recommended that you complete *your* Black Book in conjunction with '*I am Your Tennis Coaching Guru*' – this does, however, require a coach and/or parent to remain involved as it is the ultimate foundational guide towards a Top 10 tennis ranking *before* the 8 Keys were released. In contrast, if you're inside your *second* decade of play and have become familiar with the '*What is Your Game Missing*' series, then both '*The 7 Keys to Optimise Your Life*' and '*How to Develop a Top 10 Tennis Ranking – the Power of the 8th Key*' will lead you home. Nonetheless, irrespective of your PoI *your* Black Book does not discriminate against your current Level and/or ranking. What *your* Black Book does do is continuously prompt you to complete the various sections whilst it is your choice if you fast-track or opt to read twice over to ensure the embedded lessons are retained and in turn integrated into your game.

Finally, this Black Book may be your very *first* or your *third* – dependent on your Level *and* rate of progress. You may choose to continue along Level 1 for two seasons before progressing to Level 2 and then Level 3, or you may choose to complete all Levels inside the same season – as always, the choice is yours. There is no right or wrong, but caution should be taken to consider your *baseline* and your current trajectory to ensure rapid-fire pitfalls are avoided whilst progressive peak performances are favoured. After all, this is *your* Black Book and it is designed to be kept with you – from Club Level towards the Top 10 and all Levels that fall between. Whether you use *your* Black Book for one season or more, it is your superpower – *Golden* or otherwise, to accompany you towards the pinnacle of play, and to know how to stay.

Features of Your Black Book

Your *Black Book* traverses the span of the entire *Collection* in one form or another as it pulls upon this breadth of work to allow you to *navigate* the extensive terrain towards the top of the game. And it is no mean feat. Built over the course of *twelve* Blocks that contain *four* weekly *plans* to complete, **My [Tennis] Black Book** spans across the length of an entire season and then some – *forty-eight* weeks are included with *additional* days per month built into the annual *plan* – *four* weeks in total dispersed across the *twelve* month timespan.

Your Black Book affords you to choose the structure that works best for you and your current schedule and/or training load. Whilst *forty-eight* weeks are embedded in **My [Tennis] Black Book**, *not* all these weeks need to comprise of equal intensity and/or rigorous training periods throughout the season. It is up to you to choose how these weeks are best used to optimise your current training schedule and/or how you can use *your* Black Book to better *plan* for peak performance periods across the course of the season. For players ranked inside the Top 100 on the WTA or ATP Tours, each one of the *four* Grand Slams is a prime period for your performance to *peak* and as an example, *your* Black Book would structure your training accordingly leading up to each of the *four* Grand Slams followed by additional *rest* periods incorporated into your training schedule. These *rest* periods are equally important to consider and integrate as are the various periodic cycles throughout the season when you will intentionally *plan* to *peak.*

Irrespective if you are ranked inside the Top 100, Top 10 or playing at a *local* or *national* Level, the example of the *four* Grand Slams per season remains equally applicable to your own training schedule and *plan* to peak at *key* tournaments throughout the season. You may choose to work towards *one* peak over the course of an *eight*-week period, for example, followed by *two* weeks of competition then *one* week of *rest* – that's an *eleven* week period and falls slightly short of *three* full Blocks. Repeat this cycle *four* times per season and that's *four* peak performance periods achieved inside the *one* season. But this pattern or cycle will rarely be the case throughout *your* season. What's important to remember is that whilst *your* Black Book comprises of Blocks built on *four-week* periods your cycles may fall between Blocks or coincide with the beginning of your *next* Block. These Blocks are designed to manage your season across *twelve* distinct Blocks and each of these Blocks comes with its own provisions. Examples of *how* you may choose to structure your season across the *twelve* embedded Blocks is shared in Table 1. Whilst the *four* examples shared are by no means exhaustive, they serve the purpose to illustrate *how* your *forty-eight* weeks may be broken down throughout the duration of a season and *how* you can *plan* for peak periods. Similarly, the length of these Blocks are mere guides to show *how* your weeks and in turn Blocks can be used in the lead up to tournaments over the course of a season. Although the length of each of the *four* examples vary, Table 1 should *only* be used as a guide to help you structure *your* own tournament *plan* and in turn, each of your *twelve* Blocks as you build towards your respective *peak* performance periods several times throughout the season.

At the beginning of the season you will be prompted to name all *key* tournaments you will play – whether main draw or qualifying, and these tournaments will help frame your schedule whilst providing an indication of *when* your *peak* performance periods should fall. It is important to remember that *more* is not necessarily *best* when it comes to tournaments as you need to factor in *rest* each Block, per design, and controlled training periods that do *not* double-up as tournament weeks. This is where many players fall into the trap of over-training and *your* Black Book is one of many preventative measures – now a part of the *Collection* designed to steer you towards the Top 10 *with* longevity on your side. As such, *one* tournament every week of your *Black Book* is not encouraged for these reasons (as a *minimum*).

Your tournament schedule will be individual to *you* as will be your goals for each tournament. Depending on your *stage* and Level will depend on what tournaments you play and if you have *required* tournaments you need to play throughout the season. For the *required* tournaments, these will take precedence for each of *your* training periods and provide the foundation for your peak performance periods throughout the season. For those that are *not* required but are played to acquire points to *push* your ranking progressively closer towards the Top 10, whilst these tournaments are *key*, they also need to be managed across each of your *twelve* Blocks *plus* your periods of rest. Regardless of *your* ranking, this is a lot for any player to manage let alone balance over the course of a season whilst keeping on top of *your* training and *your* ability to manage *your* weekly load. This is a *key* reason behind the design of **My [Tennis] Black**

Book as it has been built to help you throughout the entirety of a season as your ever-present sidekick.

Blocks 1-2	7 Wk Period	1 Wk Rest	4 Wk Period	Tourn. (1 Wk)	2 Wk Period	Tourn. (1 Wk)	2 Wk Period	Tourn. (1 Wk)
Blocks 2-3	3 Wk Period	Tourn. (1 Wk)	2 Wk Period		3 Wk Period	1 Wk Rest	3 Wk Period	1 Wk Rest
Blocks 3-4	3 Wk Period	1 Wk Rest	4 Wk Period	1 Wk Rest	2 Wk Period	Tourn. (1 Wk)	2 Wk Period	Tourn. (1 Wk)
Blocks 4-6	4 Wk Period	Tourn. (1 Wk)	3 Wk Period	Tourn. (1 Wk)	5 Wk Period	1 Wk Rest	3 Wk Period	1 Wk Rest
Blocks 5-7	3 Wk Period	2 Wk Rest	5 Wk Period	1 Wk Rest	2 Wk Period	Tourn. (1 Wk)	2 Wk Period	Tourn. (1 Wk)
Blocks 6-10	8 Wk Period	1 Wk Rest	3 Wk Period	1 Wk Rest	5 Wk Period	1 Wk Rest	3 Wk Period	1 Wk Rest
Blocks 6-10	2 Wk Period	Tourn. (1 Wk)	3 Wk Period	Tourn. (1 Wk)	2 Wk Period	Tourn. (1 Wk)	2 Wk Period	Tourn. (1 Wk)
Blocks 7-12	4 Wk Period	Tourn. (1 Wk)	5 Wk Period	1 Wk Rest	5 Wk Period	1 Wk Rest	3 Wk Period	1 Wk Rest
Blocks 7-12	4 Wk Period	1 Wk Rest	2 Wk Period	Tourn. (1 Wk)	2 Wk Period	Tourn. (1 Wk)	2 Wk Period	Tourn. (1 Wk)
Blocks 8-11			4 Wk Period		5 Wk Period	1 Wk Rest	3 Wk Period	1 Wk Rest
Blocks 10-12			3 Wk Period	Tourn. (1 Wk)	2 Wk Period	Tourn. (1 Wk)	2 Wk Period	Tourn. (1 Wk)

Table 1: Examples to consider for Block Structure per Season

As not all weeks are mandated, each week entails different content to review. This content is in both written and auditory form whilst the pre-designed *plans* are visual and ready to complete. This three-fold approach allows *you* at least one weekly task that is preferential – that comes easier than others, but throughout the duration of the season *all* aspect of the three-fold approach will become habitual behaviours that ultimately become embedded in your overall development. *Your* Black Book has been designed to empower and encourage you as you transition from *one* Block to the next, or *one* Level to the next.

You have the option to complete *one* Level per week or all *three*. It is recommended that depending on your stage of development that you adhere to *one* Level if this is the *first* time you are using **My [Tennis] Black Book**. If you're familiar with the *Collection,* then by all means go ahead and *navigate* all *three* Levels should you choose. Irrespective of what your decision, this is *your* Black Book and if you're inside your *first* decade of play then *all* three Levels will in fact span across this timeframe and there's no need for fast-tracking. That said, if you're inside the Top 200 on the WTA or ATP Tours and are in fact on an active *positive* trajectory, then use *all* Levels at your discretion but *no* Level (or *stage*) is designed to be missed. All Levels hold equal weight in *your* progression towards the Top 10 and Level *one* forms the backbone for what's to come. What's fundamental to consider is *your* Level and stage of development. **My [Tennis] Black Book** may span three seasons or more, but it is your choice *how* you choose to *navigate your* Black Book – with *all* Levels becoming second-knowledge.

Each Level represents predetermined sections of the *Collection* to learn and/or revisit each week. As the *Collection* spans 11 Books (at this point in time), the Levels are a means of *navigating* the respective Chapters throughout the individual texts and the corresponding *stages*. For **My [Tennis] Black Book** '*Volume 1*' all Chapters shared have been refined to *stage* **one** of each of the three Levels. Each of these Levels is formed by the initial Book of each stage. By breaking down the *Collection* to span across *three* stages allows you to focus on *one* text per season per Level – no more than *three* across all Levels per season. Alternatively, *you* may choose to use the initial Book of each *stage* as a trajectory for the next per design. This design allows you to complete the initial Book of stage *one* per season for your chosen Level with all *three* Books featured at stage *one* of each Level per season inside *your* Black Book. Table 2 outlines all Books of the *Collection* and how they have been refined per their respective stage and Level.

By no means are *all* Chapters shared across the initial season of *your* Black Book but as 'Volume 1' of **My [Tennis] Black Book** suggests, stage *one* of each Level takes centre stage over the course of the 'founding' season. Whilst you may choose to read the entire *Collection* over the course of the initial season, *your* Black Book is a trajectory for your current *stage* of development – whether Top 100 on the WTA or ATP Tours, or Top 10 at your local Academy. This Level of performance denotes whether you're at *stage* one, two or three but irrespective of ranking *all* Books prior to your Level are fundamental to the next i.e. **no** shortcuts. If you are inside the Top 20, for example, all Books prior to '*How to Develop a Top 10 Tennis Ranking*' remain as *key* readings to get the most out of this text. The same applies for those of you who choose to start on Level 2 at *stage* one – *all* Books from Level 1 are prerequisites.

	Level 1	Level 2	Level 3
Stage One	The Secrets to Optimal Performance Success (2016a)	What is Your Game Missing? (2019a)	The Science of Elite Performance (2017)
Stage Two	The Secrets to Optimal Wellbeing (2016b)	What is Your Game Missing, Now? (2019b)	The 7 Keys to Optimise Your Life (2021)
Stage Three	The Secrets to Optimal Coaching Success (2018a)	What is Your Game Missing, to Win? (2020a)	How to Develop a Top 10 Tennis Ranking (2023)
Additional: The Elite Research Method (2018b), I am Your Tennis Coaching Guru (2020b)			

Table 2: Books per Stage per Level

What's important to remember about *your* Black Book is that it is *yours*. Yes, each Level contains *three* Books each but if you choose to start on Level 3 and begin with '*The Science of Elite Performance*' then over the course of the season *all* prior Books are encouraged to be read or alternatively, once you complete '*The Science of Elite Performance*' progressively tick off *all* other texts before circling back to '*The Science of Elite Performance*' for better understanding. There are various approaches to consider and whilst there is no right or wrong the only factor that needs to be considered in these approaches is that all *preceding* Books are covered. How you choose to go about this is your choice to make and to take comfort in knowing that *your* Black Book is your trajectory, as it is also up to you to consider what *stage* you are currently at in your overall development and whether the entire *Collection* is to be consumed in the space of one season or perhaps *ten* or *more*.

At the completion of your first season with the additional readings from your respective *Level* – one, two or three of stage *one* completed, you will have covered a variety of Chapters, if not *all* in the Book of focus and your game, having integrated these new *learnings*, will have reached its *next* baseline and its *next* peak performance *twelve* or more times across the span of the season. And *your* Black Book is designed to do this for *three* or more seasons before *Volume 1* comes to an end. Now that's something to get excited about.

Every *first* and *third* week of each Block will be additional **Topics** made available. Readings that will vary in **Topic**, are designed to provoke thought, further considerations and/or act as extension of the *weekly* **Chapter** in their own way. The topics are not necessarily chronologically displayed but form an initial baseline of *key* learnings to recall before gradually progressing to the 2024 season on both the WTA and ATP Tours. The later readings are in somewhat of a chronological order as WTA and ATP Tour player performances are discussed with *key* learnings interwoven throughout. When reading these **Topics** it is encouraged to review the 2024 season as and/or when indicated to afford an even firmer grasp on the *how* behind a Top 10 tennis ranking and *how* the players throughout the 2024 season fared. These **Topics** form an invaluable building block of *your* Black Book to showcase and extrapolate the resounding *how* behind respective performances with powerful lessons to again be learned.

Episodes are the next additional feature each week. These **Episodes**, directly pulled from the adjoining Podcast – Beyond Top 10 Tennis, are readily accessible per your preferred streaming platform. The **Episodes** of focus are drawn from the discussions around '*The Secrets to Optimal Performance Success*' that feature important lessons, points of recall to topics extracted directly from the Book which forms the baseline for not merely later discussions, but the entire *Collection* at hand. It is important to note, however, that whilst '*The Secrets to Optimal Performance Success*' is featured under Level *one* and is also the Book at stage *one*, it forms the backbone of the *Collection* and acts as the initial stepping stone. As such, each Book that has come *after* has built on this work and why each of these episodes offer a more detailed discussion around the set **Topic** which are also, at times, compounded by aligning with the additional weekly **Topic**. This cross-back to crossover approach ensures no stone is left unturned whilst covering various stages *and* Levels in conjunction with the core **Chapters** of '*The Secrets to Optimal Performance Success*' that also includes extracts from *'The Secrets to Optimal Coaching Success'* and later WTA and ATP Tour player rankings and results. This comprehensive *threefold* approach is designed to accommodate various *stages* and Levels whilst providing a glimpse into what's to come the further you progress.

Another feature embedded into *your* Black Book is the absence of *four* weeks from the duration of the season. Whilst the *twelve* Blocks incorporate *four* weekly *plans*, these *four* weeks are not included in *your* Blocks but can be gauged to use at the beginning and/or end of each Block. Depending on the month, these days will range from *two* to *three* days of encouraged rest – in addition to your weekly cycle that should already incorporate rest days. February will be the month of exception where you may get no additional days but will have *three* additional days in both January and

March which can be partially assigned to February i.e. two days of additional rest *across* these three months. These days also serve as a reminder to use for *review* at the beginning and end of each Block which is another feature to keep you on track at the end *and* at the beginning of each Block.

It is important to remember, nonetheless, that whilst **My [Tennis] Black Book** traverses *forty-eight* weeks of *reading*, *listening* and *planning* with *four* additional weeks dispersed throughout the season for additional rest *and* review, you may choose to take additional weeks off as you please and/or as your schedule demands. It is encouraged, nonetheless, to complete your weekly tasks during these periods of rest *minus* the training and/or performance elements of *your planner*. Other elements of your *planner can* in fact be completed during your additional rest weeks even if these are impartial as this will help *maintain* your *new* habitual behaviour.

Of course, there are other considerations for additional time off from training and/or play that may *not* be planned and/or not due to your schedule. Whether you're under the weather one week or need to take a few precautionary weeks off due to a sustained injury, use this time to double-down on your weekly tasks in contrast to letting *your* Black Book slip to the side. These unplanned periods of downtime allow additional time for reading – whether from your *current* Level or you decide to read ahead (or go back), additional time for listening to catch up on the latest insights on Beyond Top 10 Tennis, or even use to reflect on your planning and/or notes to be even *more* prepared when you're back in action.

A final note on the features in *your* Black Book is about time. Each **Episode** will vary slightly in length but those featured in **My [Tennis] Black Book** fall anywhere between 30 to 50 minutes with *newer* episodes typically 60 minutes or longer. Depending on your listening frequency and interest in taking notes, allow for a *minimum* of 60 minutes per week. The assigned Chapters are intentionally *not* too long in length as these are coupled with the additional readings every *second* week. Allow a *minimum* of 30 minutes of reading per week for Chapters and 30 minutes every *second* week for Topics. And if you're starting at stage *one* of a Level *other* than one, it is recommended that you assign a *minimum* of 60 minutes of additional reading per week to steadily chip away at the additional Books across the *Collection*. Of course, time permitting, 60 minutes **per day** would be ideal, however, this is more feasible for those of you who are already *post* formal education and without a mandate for specific study requirements i.e. unrestricted time. As for your planer, 2 hours per week as a *minimum*, should be put aside to ensure these tasks are met, however, this time needs to be dispersed across your week. An example of time distributed across *your* Black Book per week is shared in Table 3.

	Monday	Tuesday	Wednesday	Thursday	Friday	Saturday	Sunday
Chapter	15 minutes (optional)		15 minutes (optional)		15 minutes		15 minutes
Episode	15 minutes (optional)	20 minutes	15 minutes (optional)	20 minutes	15 minutes (optional)	20 minutes	
Topic	15 minutes		15 minutes		15 minutes (optional)		15 minutes (optional)
Planner	20 minutes + optional time	20 minutes + optional time	20 minutes + optional time	20 minutes + optional time	20 minutes + optional time	20 minutes + optional time	20 minutes + optional time
Additional reading	20 Minutes + optional time	20 minutes + optional time	20 minutes + optional time	20 minutes + optional time	20 minutes + optional time	20 minutes + optional time	20 minutes + optional time
Total:	55 minutes *minimum*	60 minutes *minimum*	55 minutes *minimum*	60 minutes *minimum*	55 minutes *minimum*	60 minutes *minimum*	55 minutes *minimum*

Table 3: Weekly Time Commitment

Of course, Table 3 is simply an example of how your time can be dispersed each week. The initial numbers denote a player with limited time (i.e. formal education requirements), whilst the latter numbers are for those of you with unrestricted time. It should be apparent by now that **My [Tennis] Black Book** has been designed for *you* and filled with content that spans the duration of an entire year – a full season that you can interact with and use to your advantage. Due to the variety of *knowledge* incorporated within the bounds of *your* Black Book it is suggested that your highlighter and *pen* and/or *pencil* are thoroughly used to the point of multiple replacements needed over the course of the season. When listening to an **Episode**, take notes of important points of recall and/or highlight the segment directly in '*The Secrets to Optimal Performance Success*' or *'The Secrets to Optimal Coaching Success'*, irrespective if you have the hard and/or digital copy on hand. When reading a **Chapter** from a Book from the *Collection*, regardless if you're reading from stage *one* or another, be sure to highlight *key* points to recall but also note key areas you may be unfamiliar with due to still needing to read various prior texts. And that's okay. This *should* happen as reading all 11 Books from the *Collection* won't happen overnight, rather, take your time to ensure you can transfer the *key* learnings into your game and highlight and/or note down areas you're still unfamiliar with to ensure when you do read about these 'areas' you can then refer to the original note and respective section for better clarity.

At the end of the season, *your* Black Book should be saturated with personal markings and your own performance *secrets*. After all, **My [Tennis] Black Book** is not called a *Black Book* for nothing – it is your superpower as you embark on that next peak performance, whether that is progressing towards the next-generation of play or progress *this* season towards the Top 10, *your* Black Book is your sidekick that is enriched with the *secrets* plus the data that leads the way.

Weekly Planner Explained

Your Weekly Planner is a fundamental part of *your* Black Book that could be considered one part of the *glue* that keeps all its interwoven features together. The other part, of course, is the interspersed readings of various Topics that in part run parallel with the 2024 WTA Tour *and* ATP Tour seasons, sandwiched between *your* weekly tasks and featured in weeks *one* and *three* of each Block. The planners, however, are made available to complete *every* week over the course of all *twelve* Blocks and are featured at the end of each Block. With *forty-eight* of these planners for you to complete – *four* each Block, these remain uniform throughout **My [Tennis] Black Book** to ensure an equal Level of consistency is integrated into *your* Black Book to keep you focused on the Pathway ahead. In other words, these planners do *not* change per design to ensure a Level of expectancy coexists with your current routine and/or your training load and remain as a constant for your review and reflection *each* week of *each* Block throughout the season.

What's fundamental to remember about *your* Black Book is that it is *yours* and works with you at *your* Level. For some, there will be readings that appear outside your current Level whilst for others, readings that reaffirm *known* knowledge. Irrespective where you fall, press on. These readings alike each *plan*, offer points of recall and of refresh whilst acting as a prompt to consider *your* stage and if it is time to move on to the next, or not quite. Due to the multifaceted design that ensures various Levels are accommodated in contrast to being left behind, that's where the *Collection* comes into play and only once the entirety of this *library* of sorts has become retained *knowledge* will these planners take on a different meaning. But then again, they will predominantly remain. This point of difference is shared to highlight that irrespective of your Level – Top 500, Top 10 or somewhere along the Pathway, these planners have been designed to work with *you* and *your* Level. Of course, there will be *personal* gaps due to all players being **individual** – including you. No two players will be identical in *how* they think and/or approach a given reading, **Episode** and/or *plan*, although these variances – slight or otherwise, often transfer on court during play. And that is an important note to remember. For those of you who find there is a *personal* gap you would like to include in your *planner*, room has been made available for you to add each week and also at greater length at the completion of each Block cycle or in more elaborate detail in the adjoining *Journal*.

Your weekly *planner* has been designed to be simple and yet thorough, concise and yet with the room to provide detail. If you find there is not enough space for the length of detail you wish to provide, this too has been made available for you to add each week as well as at greater length at the completion of each Block cycle or again in even *more* detail in the adjacent *Journal*. The same applies for all other areas. Each week has *two* primary entries available for brief '**Notes**' of your choice – for quick points of recall to extend upon later in your *Journal*. These '**Notes**' are also featured at the end of each Block cycle to allow for additional room to write your thoughts and/or use as additional space as you choose. By design, these additions have again been intentional due to the scale of **My [Tennis] Black Book** and the scope of content, but also to ensure that *your* Black Book can be readily taken with you in contrast to doubling in size. This underscores why the adjoining *Journal* has been built and coexists for you to add detail in contrast to only *your* Black Book – ensuring it remains compact to take *with* you whilst the adjoining *Journal* can be filled out at home. All these aspects have been considered in the creation and ultimate development of **My [Tennis] Black Book** and is shared to afford you a window into the process behind the *plans* and whilst it could very well be *ten-fold* in scale, its scope has been refined to work with *you* over the duration of an entire season and to stay with *you* – without limits, without excuse.

Comprising of *twelve* distinct sections, each of these sections has *additional* questions and/or points to address each and every week. Not only is your *planner* designed to keep *you* on track towards a Top 10 ranking, but it is also designed to provide a point of reflection, guidance, motivation to knowledge, among others aspects, as it acts as a sounding board, point of recall to collective review. Cumulatively, *your* Black Book is incredibly versatile in its offerings and these *twelve* sections to complete each week are merely the beginning. Over time you may choose to add to these *twelve* sections and/or use the adjoining *Journal* for more room to pen down your thoughts, results to progressions and more. Whatever you choose, *your* Black Book is *your* foundation – to the *Collection*, to stage *one* and to Levels 1 through 3, and ultimately the Pathway towards a Top 10 ranking.

At the beginning of each Month which also marks the commencement of a new Block, you will find room to pen your **goal(s) you would like to achieve** this specific Month. Remember, each Block is broken down into four weeks with additional days to be used at either the beginning and/or end of each month for review. These goals are encouraged to be *achievable* within this Block cycle (micro) or may commonly be known as 'short term' goals. You are encouraged to be as specific as possible to ensure you have clarity of purpose over the following four week period. Examples are as follows:

- For a player who has noticed an increase of errors from their backhand: "my goal this month is to reduce errors from my backhand (groundstroke) by focusing on greater consistency during practice and in play. I will measure this by no more than 10 errors per session on my backhand. If I get close to 8 or 9, this will remind me to knuckle down, slow down my pace to ensure I keep consistency and accuracy on my side, in contrast to focusing on power, especially if that is the primary cause of my errors. During play, I will work on keeping my backhand consistent and if too many errors begin to occur and my opponent begins to take advantage of this, I will remain mindful of my strengths and be proactive in using my forehand to my advantage to counteract these moves. Over time I hope to balance my forehand and backhand out so my opponent cannot so easily know where my errors are more likely to come from."

- For a player who is ranked inside the Top 10 and has been battling issues around their strategic play: "my goal this month is to work on my strategic play by going over various scenarios with my coach whilst being proactive in taking the time to study my opponents – whether in person or via recent recordings, and I will then consider to use footage of former Top 10 players who have had great success with their strategic play and begin to take the time to review their tactics. This month I hope with the additional work on these elements, whilst working with my coach, I'll become more strategically aware of options I can consider implementing into my game. In addition, during this Block I would like to begin to integrate newer strategies into my game so when I next play against a fellow Top 10 player who has the winning edge over me, I'll be better prepared and more confident to come away with the *win*."

Depending on your stage *and* Level will depend on the depth and/or detail provided. These initial goals set at the commencement of each Block are separate to the individual weekly goals set out in your *planner*. As such, these Block goals can afford to be slightly bigger but remain within reach over a four week period. In contrast, weekly goals are encouraged to remain within a seven day period with *four* of these micro goals incorporated into the *one* Block with its own set, slightly bigger, goal that should collectively build towards the overarching goal(s) laid bare that you would like to accomplish this season – inside the space of a *twelve* month period but shorter than the 10 Year goal penned at the beginning of *your* Black Book.

Whilst there are ample opportunities to pen down your goals for a given week, Block and/or season, as well as the years ahead, there is also room to change or update – *modify* your goals. Whether due to increased progressions, a minor setback and/or rest periods, it is important to understand that goals do change and can change which is a completely normal part of the process. By writing down these goals at the beginning of each Block and each week, come the end of the season you will be able to reflect on *how* far you have come, the changes endured along the way – remembering that *no player is immortal*, and use this as a sounding board for the *next* season and successes that lie ahead. A powerful time to leverage the adjoining *Journal*.

Whether you have gone on to *win your* maiden Grand Slam, entered the Top 100 rankings or secured your first professional match *win*, *all* these goals are triumphs and a part of the Pathway. There is no right or wrong goal to set, is it what's right for *you* that matters most. Be sure to keep this in mind each Block and each week and if there's one thing that *your* Black Book can promise you, it is to give *you* the space and the freedom to learn how to progressively chip away at achieving a Top 10 ranking.

After you have penned down your goal of focus for the Block, you will then move to **Key Takeaways**. Under this section you will find *three* headings with their respective boxes. **Episode**, **Chapter** and **Topic** Notes are available to pen *key* touchpoints of all *three* weekly tasks. As has been shared, your *planner* is one of three forms of weekly tasks, whilst you will have additional Topics to read every *second* week. For those weeks that fall *between* Topics to be read, you may choose to use the designated box for additional notes of your choosing and/or to re-read the prior week's **Topic** and to pen down additional *key* points to consider for the current week. These *three* sections are designed for a quick and brief point of recall of the respective **Episode**, **Chapter** and **Topic** of the week (or *bi-weekly*) to allow you to readily find where this *key* information is stored and/or can be accessed again when in need.

What follows next is a more refined section for your weekly goals that prompts you to answer *two* questions: one, **what you would like to accomplish and/or improve this week**, and two, **what you are working on** – short and concise. These goals are in addition to your broader Block goal and refer to the shared micro goals that ultimately contribute towards achieving your *overarching* goal for the season. Both forms of goals, whether per Block and/or per week provide the *key* foundations for you to steadily move closer towards your primary goal for the season – broken down into incremental parts over *forty-eight* weeks and *twelve* consecutive Blocks to ultimately edge closer towards achieving your seasonal goal. Be sure to make these incremental *weekly* goals count.

Talk it Out is a weekly section of your *planner* that is centred on communication with your coach and feeds directly back to the *coach-athlete relationship* which is embedded in the 8 Keys. Refined to answer *two* questions – one, **what you want to discuss with your coach this week (and to note whether you have discussed this and the outcome)**, and two, **what you want to learn more about this week (and note whether you did in fact learn more about this)**, you are encouraged to use these questions as a sounding board to engage in discussions with your coach and whilst the *second* question may be achieved through *listening* and/or *reading*, you are encouraged to accomplish this by using a mixed method approach i.e. *read*, *listen* and engage in *discussions*. Whilst refined, **Talk it out** is centred on interactions, sharing your thoughts, questions, learnings and more to ensure these are proactively heard, discussed and/or integrated into your weekly training schedule and/or your performance during play to work with *you* to progress towards that *next* baseline and that *next* peak performance Level.

Reflection is next in line, however, there is no specific order to follow when completing your plans as each *planner* is *yours* to choose *how* you complete them whether chronologically, scattered or somewhere in between with the exception for *named* days. **Reflection** is short with *one* question: your **favourite part of the week that you are most pumped for**, and whilst this is a singular question, again you're encouraged to pen more, write more about the week ahead and/or the week that 'has' been in your *notes* – that's what they are there for and for *you* to choose how you use these sections. What's important is to use these questions as a *platform* to prompt you for more and then write away. Some weeks you'll be more inclined to write than others, and that's okay. What is *key* is for you to continue to complete the *minimum* questions and/or sections to establish *your* baseline and over the course of the season when you're in need of more room to jot down your thoughts – use the adjoining *Journal*, and remember to track back to these partially empty sections and complete *your* Black Book in conjunction with the *Journal* until both are bursting at the seams. When that time comes, you know you're on track. To emphasise the importance of **Reflection**, each Block has a section dedicated to reviewing every cycle to allow you to reflect upon *your* cumulative performances. Answering this simple question will allow you to recall the multiple highs experienced throughout one Block to the next and these will serve as *key* motivators to build upon throughout the season. Ultimately, at the end of each season you will have *forty-eight* reminders of why *your* Black Book is your superpower and how these experiences have shaped your performances throughout the season and will transfer into the *next* – another element inadvertently embedded in the 8 Keys.

One of *two* **Notes** sections follows and it is up to you how you choose to use this section. Remember, at the end of each Block additional room has been made for *more* **Notes** to be penned due to the abovementioned *balance* between scale and scope, although to add *detail* this is why the adjoining *Journal* was built – especially designed to cater for this need. For more room to write, use these sections and the *Journal* for what they have been intended for – *your* answers to thoughts and everything between. Use your **Notes** to add to *your* weekly *planner* as you choose and the sections at the end of each Block, and when you crave for more room the *Journal* will be at the ready to more readily pen your weekly tasks. After all, these sections have been designed for you whilst serving as a counter between scale and scope to keep *your* Black Book compact whilst the *Journal* is filled with the room to write down your questions, answers, *secrets* and more.

Following is your **Training Session Diary** that provides *six* boxes – **Monday**, **Tuesday**, **Wednesday**, **Thursday**, **Friday** and a combined **Saturday/Sunday** box. If your weekends are currently included in *your* primary training days in contrast to weekdays then be sure to *modify* these boxes to fit your current training schedule and/or needs. What's important about each respective box is that you take the time to write down *key* learnings and/or *key* aspects of that specific session. If you have more than one session per day, be sure to specify whether a morning and/or afternoon session to provide a clearer point of recall and use the adjoining *Journal* as needed to add more detail. These boxes can be used, over time, to track peak performances and/or optimal performance timeframes whilst also providing a point of recall for *how* you have been progressing and what has been helping and/or if regressing, what has been contributing to a lull in performance, along with additional aspects of your game. Perhaps one of the most important aspects of *your* weekly *planner*, the **Training Session Diary** will provide invaluable information over the course of the season to better track *key* performance indicators and what *you* respond to best, what is ineffective and/or when a *new* periodisation schedule 'should' commence, among many more valuable insights.

If you find that there is not enough room to fit your session notes inside your training diary be sure to use the additional room at the end of each Block provided. Should you find that your *session notes* continuously search for *more* room to pen your thoughts, you are encouraged to use the adjoining *Journal* to **My [Tennis] Black Book** as it has been especially built for this very reason. For reasons shared, this additional room is intentionally void from *your* Black Book to remain true to form. As such, the *Journal* has been built to stay at home, if you choose, and to complete your thoughts in *full* with the room to do so, in contrast to *your* Black Book – designed to accompany *you* onto the courts and pen down your initial thoughts until you sit down with the *Journal* and reflect in more detail. The *Journal* is designed to complement your weekly review by taking the time to expand on your notes, thoughts, **Training Session Diary** and *all* other elements of your **Weekly Planner**, providing an additional tool to crosscheck and/or be filled with notes and over time, your foray into the *Collection*.

Next is a section dedicated to *your* **Progress** with *four* underlying questions to reflect upon and answer on a rolling basis. **Top 3 superpower moments** refers to *your* best tennis throughout the week when you felt on top of the world. Whether a tweener that landed as intended, a serving milestone or that perfect session when you reached a *new* Level and a *new* zone was locked into place, these are moments for you to relish and track as over time *your* **superpower moments** are the ones that build that *next* baseline and progress you closer towards that Top 10 ranking. **Any goal updates you would like to make** refers to adjustments you may like to make, whether to your primary and/or overarching goal, to your Block goal, or weekly goal(s). This is a moment to reflect on *your* goals each week and to simply ask yourself if there are any changes you would like to make – additions and/or otherwise. There is no right or wrong answer when considering **any goal updates**, rather a question for *you* to review your goals to ensure they remain true for *you* and pen any thoughts, concerns or otherwise to reflect upon. **Top 3 key learnings this week to bring forward into the following week** is all about pinpointing your **Top 3** moments this week that you consider as **key learnings** that are invaluable for you to retain and continue to work on and/or integrate into your game. There will often be *more* than *three* key learnings **to bring forward into the following week** but this question is about asking yourself the *most* important *three* learnings and for any additional **key learnings** you would like to remember, be sure to add these to your *notes* and/or *Journal* as important points of recall. **Biggest motivator of the week** is an invaluable reminder of what lights that fire from within and then gives it an extra *push*. What you *choose* to write here will be instrumental each and every week, whether the same week in week out, or something different *each* week or *each* Block. There is no right or wrong answer to your **biggest motivator of the week** but simply a reminder of what can *help* in times of need, be it that extra stimulus to get you closer towards that *next* peak performance, to reach that *next* baseline by the end of a Block and/or for days when you seem to have hit a performance plateau, a gentle nudge to give your performance a boost. These responses can serve as equally powerful tools to draw upon during times of need that will vary from player to player and may also provide a source of near instantaneous motivation *during* play – be it a 1st Round encounter, a final or a qualifying round, *your* Black Book is quite literally *your* toolkit filled with resources to use to *your* advantage.

Tournaments is the next section to complete and whilst you will not have a tournament *every* week of your *forty-eight* week *planner*, you will be able to answer its *two* questions: **tournament you are playing and/or following this week**, and **favourite player to watch in-action this week**. If you are playing a tournament during a given week it is important to note this event as a point of review for your additional *notes* and *results*. If you are not playing a tournament during a given week there will undoubtedly be one you are keeping a close eye on – whether a Grand Slam, a WTA 1000 event, an ATP Masters or *any* tournament, irrespective if an ITF event or a local Club tournament. By noting what event you are following in a given week will show a *pattern*, over time, which may inadvertently influence your training and/or playing, or simply heighten your understanding of the predetermined *knowledge bank* behind the 8 Keys. Either way, keeping track of whether you are playing a tournament or *not* during a given week is an additional insight that feeds into your cumulative performances and you will be more readily able to track training peaks in alignment with peak performances *during* events. Similarly, the tournaments you choose to follow may be one you would like to play *next* season, one you *aspire* to play in the years to come and/or one you're sidelined from and are eager to keep a close eye on your competitors. Either way, the **tournament you**

are playing and/or following this week is another important point of recall to reflect upon and review at the culmination of *each* Block and *each* season. As for **your favourite player to watch in-action this week**, this is relatively self-explanatory. You may have certain players you absolutely thrive off watching, those you learn from and/or those you simply must watch every chance you get. Over time, these are the players that shape your game, whether slightly and/or more noticeably, from your favourite shots, to at times, playing quirks and more. By noting your **favourite player to watch** affords another point of recall from how your game is influenced, or to better understand one of your extrinsic motivators that can inadvertently be used, for example, to further your performance in various ways.

A section titled **Non-tennis** follows with *two* simple questions to reflect upon life *outside* of tennis. Irrespective of what Level you are at and/or what ranking you hold, **Non-tennis** is all about *you* away from tennis and what *you* enjoy – be it reading, gaming, devouring the arts and/or one of many countless **Non-tennis** hobbies. **Highlights of the week** is your chance to note the highs of *your* week and what you've most enjoyed this week – *away* from tennis. Discussed in detail through various Chapters of the *Collection*, your identity away from tennis matters just as much as your tennis identity if not *more*. After all, whilst you're playing tennis now at either a high and/or elite Level or aspire to do so, this unfortunately won't always be the case. Whether your playing career spans *four* decades, *two* decades or somewhere in between these milestones, the time will come for you to start that next chapter in your life, as is the case with *all* athletes who are fortunate enough to build their initial careers in the sport they love. And irrespective of what Level you are at, no player is immune from the day when the time comes to close a high performance and/or elite Level playing chapter. This is not to say you will stop playing tennis, but rather if you are on the WTA or ATP Tour, the time has come for your body to simply take a step back and if you choose to continue to play, it will simply be away from the WTA or ATP Tour and/or at a reduced load. Whilst this is not a scenario *now* as you work your way through **My [Tennis] Black Book**, it is important now to keep your identify – to not wholly be consumed by tennis. Rather, it is a balancing act and keeping a hobby or *multiple* that you enjoy away from tennis serves an important role *now* and in the decades to come. **Hobby you are most enjoying** may be identical each week and/or different each week, depending on how many hobbies you have – *one* or multiple. There is no right or wrong answer to *how* many hobbies are ideal, rather what you enjoy *doing* away from tennis and intentionally find the time to do *each* week and/or every *few* days and/or weeks simply because you enjoy doing it. If you would like to add to your **Non-tennis** section, go ahead and share your thoughts in the adjoining *Journal* to ensure *all* of your thoughts, answers and more, find a place to be shared each week.

Improvements includes *three* questions to recap your week and learn from as you head into the following week. **What you want to improve from this week** may relate to *your* training intensity, *your* results during play and/or an aspect of *your* technical foundations. Of course, **what you want to improve from this week** will vary from week to week and or may remain constant for the duration of an entire Block. What's important, however, is to note down what you're working on and set on improving and the more specific you can be, the better, as this will offer a powerful point of reflection in the Blocks to come, from timeframes/duration of improvement, alternative areas you want to improve and/or something else you're currently working on and want to improve over a set period of time – be it *one* week or *one* month. **Biggest distraction of the week** is an important element of your overall game to consider as like other questions, patterns may emerge and/or by simply writing the distraction down will allow *you* to be more conscious of its impact on *your* training and to consider possible solutions to lessen the known distraction over time. **How was my nutrition this week** is a singular point of review in your weekly *planner* to reflect upon *your* energy sources, what you're using as fuel and how your recovery has been tracking. It is *your* choice how detailed you choose to be when considering your nutrition over the course of the week and the aspects of your collective nutrition that are important for you. Whether brief and/or in depth, you may choose to share more in the adjoining *Journal*. The various intricacies of *your* nutrition have not been outlined to ensure you can choose what to incorporate depending on your current Level and/or age range, for example, with additional factors. **How was my nutrition this week** is also a reminder to consider its *influence* on your performance over the course of the week and serves as a prompt to evaluate your energy levels, recovery times to fuelling efficiency and more for those of you at higher levels of play of which it is anticipated that this is an area you consistently monitor. For those of you at the earlier developmental through to high performance levels of play, your nutrition will be an area you are progressively learning about over time and as such, the detail you provide is not anticipated to be as detailed as a player who is currently ranked on the WTA or ATP Tours. Due to the combined detail and distinctiveness of nutritional plans from one player to another, this is not a component of *your* Black Book although introductory points of nutrition in general are present in the *Collection* and provides a non-athlete specific resource.

Next is **My numbers for the week** which contains *three* areas of your week to enter in *numerical* form. **Hours on Court** provides a weekly recall of your playing *load* each week. This number serves as an important data point to monitor your **Hours on Court** to ensure any increases and/or decreases can be accounted for whilst serving as a preventative measure for over-training that in turn can be attributed to a heightened susceptibility to injury. **Hours on Fitness** reflects the time you have spent on your *strength and conditioning* that is not included in your **Hours on Court**. Your **Hours on Fitness** is an integral part of your training load to be monitored as alike **Hours on Court**, an increase too soon and/or if these hours are not considered in conjunction with *your* on-court training hours, you will become more susceptible to overload and ultimately at greater risk of injury. **Work Ethic / 10** is for you to rate your overall working capacity *'out of'* 10 each week. This is a measure of your overall discipline, dedication and determination – the 3D's referenced in this breadth of work. Collectively, these *three* numbers at the end of each week provide a value for you to consider and to maintain and/or steadily increase and/or decrease – pending performance requirements, whilst your work ethic is to be maintained and/or improved. Each of these numbers provides yet another point of reflection *each* week, *each* Block and over the course of the season to review, reflect upon and to help gauge performance measures over set periods of time whilst helping *plan* for the next *stage* ahead.

A section for your **Matches/Results** follows with *three* points of entry incorporated in your **Weekly Planner** plus *three* more incorporated in each end of Block Review plus *three* or more available *after* Week 4 of *most* Blocks but *before* your **Weekly Planners** available at the *end* of each Block. These additional **Matches/Results** have been designed to fit *your* Black Book and as such, some Blocks will have more **Matches/Results** available than others – Blocks 1, 2, 6, 8 and 12 have *six* or more to account for Blocks 3, 4, 7, 9 and 11 not fitting these additions whilst Blocks 5 and 10 have *three* each. In total, that's *fourteen* sets of *three* dispersed over the 12 Blocks plus your weekly *three* and end of Block *three.* Over the course of your season, this provides room to record **4** to **5** matches on average each week over the course of all *forty-eight* weeks. In the adjoining *Journal* there is room provided for *all* **Matches/Results**, however, these **4** to **5** points of entry are the only entries with the predetermined titles: **opponent (name), tournament (name), score, your strength, your weakness,**

opponent strength, opponent weakness, other (notes). Each of these titles leaves limited room for more, rather these are simple and *quick* points of entry for recall. Should you play more than **4** or **5** (depending on the Block) matches in a week, this is where the adjoining *Journal* comes into play. If you require more than the predetermined **15** to **24** entries per Block, the *Journal* provides ample room to fit *all* **Matches/Results** *and* key weekly tasks to truly build out *your* Black Book in full. What's most important about these points of entry is your responses. These entries serve as your *secrets* to use against your opponent in future matches whilst also providing you with notes on how to better prepare. This is a fundamental aspect of *your* Black Book as over time these *secrets* will amass to **two hundred and sixteen** if *all* points of entry are completed throughout *your* Black Book over the course of the season. You can further increase the power of *your* Black Book by using the *Journal* to add even more records – an additional *two* **Matches/Results** included per week, for example, over the span of each **Block** will add **ninety-six** more records – bringing your season total to **three hundred and twelve** records. That's a significant asset to use to *your* advantage over time whilst also considering **your strengths** and **your weaknesses** and how these transform over time. These entries can be used to help your performance in various ways, from your weekly training to *key* areas to improve and/or *new* considerations to work on implementing during play. How you complete your **Matches/Results** is your choice, however, the more accurate and complete these points of entry are, the more robust *your* Black Book becomes as it helps steer you towards the top echelon of play – and *stay*.

Should you not have **Matches/Results** to enter during a given week, you may choose to enter practice games and/or points. If you have the *same* opponent on more than one occasion during the week, consider mixing this up. If that is not a possibility and all *six* matches are with the *same* opponent, use this as a platform for *growth*. You will learn from one game to the next whilst getting the chance to improve on your strengths to weaknesses in a short time frame in contrast to perhaps waiting weeks or months (or longer) until you play them again. Look for the advantage and you'll find it. If you're not competing but are able to play matches inter-Club, use this as your platform to progress from. Alternatively, if matches are out then consider *observing* a number of your favourite matches during the week and prepare your entry accordingly as if *you* were playing so when the time does come and you do play against these players, you will be better prepared by having these *points of entry* in your arsenal. Of course, you may choose to use a variety of these options if you are not playing *six* or more matches in a given week and irrespective your Level of play and/or your current *stage*, these **Matches/Results** are yours and yours alone to add to and progressively build out the power of *your* Black Book. These *secrets* and all additions to *your* Black Book will help you progressively optimise your game and secure more *wins,* be it at Tour Level, Club Level or in the final of a Grand Slam. Remember, *your* Black Book sets the benchmark for what's to come.

The second of the *two* **Notes** sections follows for you to use as you choose. If you have not filled up your **Notes** by this point, review each of the sections of your weekly *planner* and consider taking a minute to add a little more. If you're struggling to find room that's a positive sign – use the *Journal* as intended to complete any additional entries. By this point, all sections have been completed and what remains is the **End of Month Block Review & Additional Notes**. Whilst you may have used this room for additional notes, the **End of Month Block Review** awaits. You may choose to be brief and/or provide in depth detail on *how* the past *four-week* Block unfolded and *how* it has impacted your performance – use the *Journal* to your advantage. There is no right or wrong but simply a gentle nudge to take the time to *review* the Block you have just completed, whether short, detailed or a semblance of both is yours to choose. What's important is to take the time at the end of each Block and to use the additional *one*, *two* or *three* days set aside at the beginning of the Block to simply review your notes, reflect on your performance and/or results and consider what you would like to bring forward to the next Block. Whether it is your **work ethic**, the upkeep of the **hobby you are most enjoying** and/or a favourite **superpower moment**, for example, these are merely points of recall to set the next foundation for your next Block as *your* Black Book gradually comes to life.

With all *twelve* sections now shared of what your **Weekly Planner** comprises of, you can see how *your* Black Book will transform into a micro library come end of season. In conjunction with the *Collection* and its extensive library of knowledge, **My [Tennis] Black Book** follows suit, however, you're in the driver's seat. Beginning with **Key Takeaways** that reinforce *key* learnings read and/or heard, this sets the platform for integrations into your game for the week ahead and/or *key* points to be mindful of as you tackle the following week. **Goals** ensure your weekly intent remains clear, whilst **Talk it Out** is pivotal in ensuring you hold up your end of the *coach-athlete relationship*. **Reflection** offers a positive twist of what you're looking forward to in the week ahead, before **Notes** offers room for any additional thoughts. **Training Session Diary** follows and is instrumental in monitoring your daily performance whilst **Progress** prompts you to *reflect* on your week and pen your highs to learnings in *three* and then more. But all these sections are short and concise as the adjoining *Journal* is built for *more* – to capture your thoughts, answers to reflections in *full* as they begin to overflow.

Not necessarily in order, your **Weekly Planner** is to be completed as you choose. Whilst some elements need to be completed prior to the beginning of the week and others at the end of the week, there are then sections that fall daily and/or in between. As with each aspect of *your* Black Book it is your choice and what works best for *you*. **Tournaments** feature next whilst **Matches/Results** are featured near last – *before* the second section for **Notes**. Whilst there may be a crossover between **Tournaments** and **Matches/Results** this mightn't always be the case if your matches fall shy in a week. Both sections are integral, however, in positioning your weekly **Matches/Results** and integrating your *secrets* that will transform *your* Black Book into its namesake. **Non-tennis** follows with an important reminder of identity before **Improvements** offer another point to reflect. **My numbers for the week** is the last of the sections, although not named chronologically, forms a vital piece of your *planner* by equally weighing performance load whilst mitigating injury providing your training remains progressive *and* includes the parameters of the 8 Keys.

Ultimately, your *planner* is *yours* to interpret and complete whilst the abovementioned criteria is *your* guide. The various aspects of your *stage* and your respective Level will in turn shape your output – dependent on what **Chapter** you are reading and/or *if* you are reading an additional **Topic** that week, these features of *your* Black Book add to your knowledge bank each week. As with each **Episode**, the *key* learnings that resonate with you will influence your *plans* due to the multifaceted nature of **My [Tennis] Black Book** and collectively, each of these features serves its own purpose in imparting how to *navigate* and *plan* towards a Top 10 ranking.

Tournaments to Play and Goals

Date	Tournament	Goal Result	Date	Tournament	Goal Result
Jan			Jul		
Jan			Jul		
Jan			Jul		
Feb			Aug		
Feb			Aug		
Feb			Aug		
Mar			Sep		
Mar			Sep		
Mar			Sep		
Apr			Oct		
Apr			Oct		
Apr			Oct		
May			Nov		
May			Nov		
May			Nov		
Jun			Dec		
Jun			Dec		
Jun			Dec		

It's Time

How to *navigate* and *plan* towards a Top 10 ranking has never been done before. This cannot be stressed enough. **My [Tennis] Black Book** has been built for the next generation of play – whether this season or the *decade* to come, giving you the tools to follow the Pathway and progressively edge closer towards the top echelon of play. It is also fundamental to reiterate how science has been used to build out the *Collection* – your on-call library to work your way through per your respective *stage* and Level.

Science is unfortunately often misunderstood. The science involved in building out the *Collection* through to *your* Black Back has comprised of various disciplines and lengthy studies on players *and* coaches on the WTA and ATP Tours. These datasets have contributed to *all* previous Books to come before **My [Tennis] Black Book** and underscores why they are instrumental in this process. Over time, you will work your way through the *Collection* with *your* Black Book in hand and progressively learn how science and all *key* data-points on the **best players in the world** over the past decade *plus* has allowed the 'how' behind *achieving* a Top 10 ranking be brought to life.

Each week you will embark on a new *plan* with new **Key Takeaways** to pen, **Goals** to write and **Progress** to share. You will be prompted to **Reflect** and **Talk it Out** all whilst keeping pace with your **Training Session Diary**. You will be reminded to take time for **Non-tennis** activities and to write down your **Improvements** whilst tracking your **Numbers for the week**. You will also have room to record your **Matches/Results** each week along with **Tournaments** you are either playing and/or following plus *more*.

Each week you will be presented with a new **Chapter** (or *two*) to divulge, a new **Episode** (or *two*) to tune into, and every *first* and *third* week, a new **Topic** to reflect upon. You will also have the choice to complete one primary *stage* and/or progress through more than one Level. This is for you to choose, however, if you are unfamiliar with the *Collection* it is encouraged that you begin at *stage* one and the corresponding Level and as shared in Table 3, use this allotted time to read ahead of your current Level should you choose. For those of you who are in the initial phase of your *first* decade of play – take your time from stage *one* through *three*. For those of you who are in your *second* decade of play and are either embarking on a *professional* career or currently hold a ranking on the WTA or ATP Tours – Top 500 through to the Top 50, work with Table 3 to use this time to *your* advantage to keep pace with The Long Game. And for those of you ranked inside the Top 50, Level 3 is for you, however, Level 2 is a prerequisite and Level 1 forms the baseline. As such, you are encouraged to increase your allotted time over the course of the season to prepare to integrate the *key* learnings and *all* of the 8 Keys into your game to become a barrier breaker the following season, if not sooner. Although, if you *are* ranked inside the Top 10 and are in search of your maiden Grand Slam or perhaps eager to *add* to your tally, Levels 2 and 3 are a must – there are no shortcuts, yet finetuning the integration of *all* 8 Keys at a higher rate of frequency will position you to claim the *win*.

And now, this is from me to you. **My [Tennis] Black Book** has been built for you – *from* me, to you. As I cannot be there in person for each one of you, I had to think long and hard about *how* I could still be there to help you reach your *full* tennis potential. Keeping this in mind, I have designed *your* Black Book for you if I was *your* Coach with the shared objective of achieving a Top 10 ranking. From all my *research*, *publications*, *episodes* and *more* – these have become embedded in **My [Tennis] Black Book** to accompany you along the way whilst serving to help you *navigate* through this body of work.

Your weekly *planner* is a tool that I genuinely want to see. Collectively, your *plans* in conjunction with the *Journal* forms the backbone of *your* Black Book and builds a custom *dual* **Black Book** for *you*. The *secrets* you will pen over the course of a season cannot be undervalued as they afford a comprehensive window into *your* performance for *your* Coach to see. And that's another addition of **My [Tennis] Black Book** in that it serves *two* – also designed to help *your* Coach help you integrate the core pillars of the *Collection* whilst also enhancing their own coaching pedagogy should they aspire to follow the Pathway with you each step of the way.

Be warned, however, that you will have harder weeks throughout the season but if you stay true to *your* Black Book it will get you through. Each **Chapter** and each **Episode** to **Topic** serves its own purpose in *navigating* the Pathway whilst equipping you with the knowledge to go one step further. Over the course of *your* Black Book each *secret* will become its own contribution to your growing knowledge bank to be valued, as collectively each *secret* works in unison to establish that *next* baseline and that *next* peak performance. And before you know it, after innumerable baselines formed *and* peak performance cycles, *navigating* towards the Top 10 is all within reach. Stay consistent, stay disciplined and you will steadily gain the advantage, with time, as the **next generation** of play integrates itself into your own game.

Take caution in knowing the 8% exists for a reason *or* the 0.8% – depending on your ranking *and* perspective. But take comfort in knowing that what has never been done before is yours to embrace with a significant first – having *science* on your side. In time, this will take on a more profound meaning and you will equally learn to grasp the depth of the work behind *your* Black Book and how *navigating* towards a Top 10 ranking is now between the palm of your hands. On that note *your* Black Book awaits to begin, from your very first Block and until all *twelve* appear to be bursting at the seams. Only then, will I see you for additional remarks and one final lesson, enfolded between the pages.

January: Block 1

GOALS:

Week 1 of Block 1 Tasks

Episodes to listen to: #2 *and* #13
Book Chapters:
Level 1: *The Secrets to Optimal Performance Success (2016)* – 1.2 *and* 2.6
Level 2: *What is Your Game Missing? (2019a)* – Ranking 1 *and* 1
Level 3: *The Science of Elite Performance (2017)* – Chapter 1: Anticipation / Perception

Topic: Hiring and Firing Your Coach, Rate of Progress and Balance

One of the most controversial topics out there but perhaps the most empowering, is putting the control back into the players and/or athletes hands to choose *who* they resonate with best and/or *who* they would like to have on their team. Changing this perspective can be an incredibly rewarding experience for a player as the preconceived roles are flipped on their heads giving the player the power to really think about 'who' they'd consider as an invaluable addition to their playing future.

Too often coaches manipulate their players into believing they are equipped to deliver their lofty dreams and/or aspirations without any prior experience or more importantly, education in really being able to help the player understand the road ahead and what steps need to be taken over set periods of time. But this is an important Topic and one that deserves cause for concern. To begin, there is zero correlation between a former Tour Level player and their ability to be a 'good' coach let alone promise Grand Slam success. When considering historical performance, per the 8% discussed in *'How to Develop a Top 10 Tennis Ranking'*, there is a limited number of Tour coaches who have been able to achieve Grand Slam success with each of their players. Sure, there may be a handful to do so with a select few, but very rarely has a coach been able to take a player outside the Top 50 (let alone Top 100) and progress them towards the Top 10 and then do this with another player. There are more common occurrences when a player is already ranked inside the Top 20, for example, and for a coach with a fellow Top 20 player to switch teams but remain within a similar ranking range. For some, this works and for others it does not. Either way, there is no guarantee unless the 8 Keys coexist and when this has been and/or is the case, this is when a *new* Grand Slam contender *or* winner is crowned. These are the coaches who are privy to the 8 Keys and in turn, develop them in their players (or the *Keys* missing in their game) to achieve this Level of success.

> But manipulation is real. If the 8 Keys are not being discussed then be warned. There's a whole world of 'specificity' out there and there are also a lot of tools to be leaned on to help upskill and/or *navigate* the entire coaching process.

For the player, one of the most significant aspects of *hiring* a coach is the awareness around what skills do *you* need to learn, at this point in time, to progress further — and the type of coach that is fit-for-purpose. This may or may not be a coach who'll be by your side for the next 10-20 years, or for the next 6 months — but hopefully once you come to learn what works best for you and your performance the *firing* your coach becomes less frequent. When you do come across a coach who is open to embracing the 8 Keys or already has these in their coaching pedagogy, you'll know you're on the right track. But this is not the best fit for everyone.

Some players will prefer a staggered approach — developmental to high performance before elite, with various coaches along the way. It is incredibly important to note, however, that this is *not* correlated with a Top 10 ranking, nor does it coexist with the 8 Keys. Irrespective of how many coaches along the way, if they are not privy to the 8 Keys you'll remain on the backfoot. Of course, changes in coaches is fundamental until you find that right fit and that's what it is about –finding what works for you and is conducive for your performance.

It's incredibly important to choose wisely and that is where the parent/guardian comes into play to help you look through a new lens and gauge the perspective required to really think about the journey ahead. Although truth be told, most parents/guardians will listen intently to each coach in contrast to asking these questions around the 8 Keys. And that's why the *Collection* is instrumental for **My [Tennis] Black Book** to help you and your parent/guardian (as applicable) *navigate* the way.

Of course, an incredibly controversial and yet important discussion to be had between player and parent/guardian *before* between player and coach, the discussion around Hiring and Firing Your Coach should be empowering. After all, the player is in control and if it is your dream to one day *win* a Grand Slam or to *win* multiple Grand Slams, this can be a make or break discussion irrespective of what Level of play you're at – Top 20 on the WTA or ATP Tours, or Top 20 at your local Academy, this is a pivotal discussion and lesson for all. And if you are one of these players, embrace the discussion and feel empowered by your actions to initiate these conversations with your coach and if they are not being upfront and tackling the hard questions with you, then it is time for them to step aside.

When considering a players rate of progress there is an important element that is often overlooked – timing. Collectively, the sporting world is remarkable with various levels to dynamics at play on and off the court across a multitude of sports – it is quite phenomenal. Unfortunately, of the millions of children/adolescents/young adults that play a sport across the globe the rate of failure is remarkably high. More players stop playing than there are those that continue to play at a high Level due to a number of reasons – from availability, accessibility, injury to gender disadvantages and more.

For the persistent player and the ones who are lucky to escape injury (or one that is yet to stop them in their tracks), these are the ones to overcome the multitude of hurdles in front of them before they even start to progress towards the elite ranks but these are the players ahead of the millions worldwide who never got the chance to progress to these heights. Of course, this is not an absolute number due to the immense numbers of sports participants globally and quite simply, there's no precise record due to the various causes that can be attributed to a player's rate of failure. But what's important here is that this 'failure' is not necessarily due to the player themselves or their Level of play, but because something and/or someone derailed their progress and as such only a limited number of players have a real chance – ahead of millions of others around the world to follow the Pathway unobtrusively.

But there is no guarantee. Nor is there a distinct ratio between performance and timing. And that's the hard truth that is often ignored by those in positions of power (i.e. governing bodies) that turn their backs on aspiring players all because they are a year or two over a certain benchmark that in reality, doesn't really exist at all — forming *another* barrier. Due to this outdated belief, countless players who have been progressing are blocked from the support to go further. And you're not alone. This has been happening for a long time.

A direct consequence of this timing and a player's rate of progress is the cost factor that is inexplicitly tied to these results. And the costs behind progressing from a junior to senior tennis player – embarking on a career on the WTA and/or ATP Tours, can be quite astronomical. But if you're reading this, there's good news in store for you. Whilst **My [Tennis] Black Book** is one Book in the

Collection of Books I have penned, there is a complete *Collection* that is now at your disposal and with *your* Black Book in the palm of your hands, it will help you navigate the *Collection* plus so much more – reducing these cumulative costs by 65% or *more* over the initial decade of play. Why? Because I've been in your shoes and as I shared at the very beginning of *your* Black Book, your dreams are mine, too – I want you to succeed and reach for the **stars**, too.

But a little bit of balance goes a long way. Remembering to take time for your non-sporting self is fundamental for the road ahead and why it is featured each week of **My [Tennis] Black Book**. Balance is often overlooked or simply forgotten due to outdated attitudes around 'train hard' and the 'guilt complex' if a player is not training hard enough. Unfortunately, burnout does happen and it is a real thing. Likewise, it is important to remember the failure rate or drop-off ratio whether on a local or global scale. Either way, the likelihood of a player 'breaking through' is incredibly low due to the lack and/or limitation of resources. Fortunately for you, one of those resources is in *your* hands.

Being aware of the role of balance is a healthy starting point. Implementing balance and making sure it is included in your weekly training diary and/or schedule as time for *you* is the next step and hopefully one that is up kept. Interestingly, the role of balance may very well be the deciding factor between continuing to play to walking away from the sport altogether. Irrespective of age, there is a clear reflection in mainstream 'professional' standards in establishing a work-life balance that talks directly to the role of *balance* in life.

> The irony here is that the conversation of balance has been neglected in earlier developmental years and is commonly reflected in the parents/guardian's 'professional' life and/or embedded in their belief system. And now it is time to flip the switch. Balance works for all and in dual ways.

Parents/Guardians lead by example and often these beliefs will be passed down to their children. Whether you're a teenager or young adult, you may have varying beliefs yourself around balance and if these are not conducive with your performance, it is never too late to introduce this widely known concept to ensure you have adequate rest time and time to recharge. Without this, you'll become even more susceptible to injury *if* you've managed to dodge this downfall until now.

Work-life balance is synonymous with player-life balance. The issue in question here is around encouraging a Level of balance in active players – irrespective of age whilst still flourishing in their performance. Without this Level of reprieve integrated into a player's training load, the player-life balance becomes susceptible to following in generational footsteps and as such, burnout and the opposite of longevity at the top of the game will follow – before the player has even managed to crack the Top 100 *if* they are lucky enough to get that far.

Should balance be void from a coach's pedagogy, parents/guardians may be none the wiser. But now you know better. The same applies for the discussions around rate of progress and the role of timing – be careful and use the tools herein provided and from the *Collection* to follow the Pathway. And when it comes to Hiring and Firing you Coach, the answers really are inside the 8 Keys and why each one of these Keys is absolute in its premise to deliver a Top 10 ranking.

Week 2 of Block 1 Tasks

Episode to listen to: #4
Book Chapters:
Level 1: *The Secrets to Optimal Performance Success (2016)* – 1.4
Level 2: *What is Your Game Missing? (2019a)* – Ranking 2 *and* 2
Level 3: *The Science of Elite Performance (2017)* – Chapter 1: Athlete Communication

Week 3 of Block 1 Tasks

Episodes to listen to: #17 *and* #18
Book Chapters:
Level 1: *The Secrets to Optimal Performance Success (2016)* – 3.1 *and* 3.2
Level 2: *What is Your Game Missing? (2019a)* – Ranking 3 *and* 3
Level 3: *The Science of Elite Performance (2017)* – Chapter 1: Biomechanics / Technique

Topic: Functional Movement – Centre of Gravity and Range of Movement

Too often a tennis player's Centre of Gravity (C.O.G) is overlooked and not accounted for when developing the overall technical foundations of their game. Irrespective what Level is considered, there is often an unbalanced C.O.G that results in a poor frame and the player needing to use their upper body to 'muscle the ball' – placing the majority of the load on the half of the body *not* responsible for producing the key biomechanical attributes that *are* attributed to a Top 10 ranking. There's a reason why a player's lower body is so fundamental and for their upper body to work in tandem in contrast to being out of alignment and/or separated. If this relationship is not built for an 'balanced' exchange, not only do injuries run rife but a player's playing days will rarely see that 10 year peak lifespan.

Although there's a slight correction. When considering that 10 year period this is in respect to 'peak' longevity at the top of the game – a Top 10 ranking with deep runs at the Grand Slam Level with the high probability of Grand Slam success. In other words, you can still play tennis for a 10 year lifespan, just *not* at the pinnacle of play with the denoted rate of success. As such, it is a rather alarming trend when players are conditioned to hit the ball without being mindful of their C.O.G.

A player's C.O.G is interchanged with their balance and their ability to maintain a Level of dynamic balance. If this is a struggle, then you're a red flag when considering a Top 10 ranking let alone one that is equipped to progress towards this ranking range. By being able to interwork your C.O.G into your technical foundations, you're equipping your 'whole' body to work in sync from the Level of control to the role of a player's C.O.G in producing power to performing under stress.

> Often developing players are encouraged to hit the ball and work on their power opposed to learning how to accommodate their C.O.G and use it to their advantage.

This is an area that is often overlooked but equally integral for preventing long-term injuries along with more fast churning ones i.e. elbow and/or shoulder related, among others. By being more conscious of how the body works and the benefit of functional movement patterns, not only is a player preparing for The Long Game but setting themselves up for a greater rate of success to progress towards a Top 10 ranking.

When considering a player's Range of Movement (R.O.M) there are similarities between a player's C.O.G but these two mechanical inferences are two entirely different concerns but do in fact work in tandem. Often overlooked and a cause of a player's weakened performance, the range in which a player moves and their capacity to move, whilst maintaining a state of dynamic balance, is controlled by their C.O.G.

This is a technical mishap that often occurs and causes a player's body to be 'overstretched' too soon and/or become more robotic when this key fundamental movement principle is not developed. Simplified, a player's R.O.M considers their capacity to reach and/or extend a part of their body to perform a certain movement and in tennis, this happens on countless occasions during a single rally. Being mindful of the body as a whole and how you move from a functional perspective whilst taking progressive steps towards your desired range, this is how a player's R.O.M is gradually built and learns how to accommodate the increasing demands as your game evolves. But too much too soon is equally a cause for concern as you may very well overextend without the strength and/or sturdiness of a more developed R.O.M whilst your C.O.G may struggle to keep you upright. Whilst this is a simple example, it serves the purpose in sharing the negative connotations associated with neglecting to develop a player's R.O.M as although a fall may be perceived as minor, in more demanding scenarios this can translate to a more serious injury due to an overextension and/or unbalanced state that can leave a player sidelined for a set period of time.

Accommodating the role of a player's R.O.M and progressively developing its use and its capacity, progressively, allows a player to better understand how their body works and why the role of their R.O.M and C.O.G are in fact significant indicators that are responsible for other functional demands, from elastic energy to the role of momentum. By taking a steady and/or progressive approach, not only are you giving the athlete body ample time to adapt and modify its movement patterns, you are also safeguarding the player body for the road ahead.

One of the most significant contributors to mitigate injury in the aspiring and/or active player is the ability to integrate functional movement patterns into their game. The ability to adopt a stance that is open to and accommodates functional movement is a key indicator and a head start that most developing players are deprived of and in turn not taught until well into their formative years when they are on the cusp of potentially turning professional. The draw back here is twofold — one, the player needs to recondition their current technique to incorporate functional movement patterns and two, they are more susceptible to developing an injury due to their repetitive motions to movements not aligning with functional movement principles. More concerningly, players who progress through their developmental stages and towards the professional ranks with these patterns being all but absent, are in fact at a higher rate of detriment and at risk of sustaining a long-term injury.

> It's incredibly important to note that many injuries sustained in the professional ranks in fact started many years prior due to these simple function movement patterns not being instilled in their game.

And with this knowledge *you* can better prepare for longevity at the top of the game, be that the Top 10 through to the Top 30, without needing a hiatus from play due to a part of your body simply no longer being able to withstand the demands of the sport. Interestingly, the role of function movement is not incorporated in coach education literature and/or coach education frameworks universally and simply explains *why* these injuries continue to happen – at the developmental through to elite levels of play. But should *you* know better, you can mitigate the associated risks by integrating these *key* foundations to sustain your Level of play now and in the years ahead as you're equipped to maintain your hold at the top of the game.

By integrating these patterns in your game, from the role of your C.O.G, R.O.M to momentum, a player is setting themselves up for success. In contrast, imagine using a technique, or multiple that do not align with functional movements patterns and respective patterns of play on repeat for ten plus years. If your body is not equipped and/or accommodating these movements in a way that works *with* your body, that's a potential ten years of overuse that is wearing down your body as you become more susceptible to injury. To avoid these types of injuries, functional movement patterns and their integration into your game is key and one of their natural by-products is momentum. By using momentum in your game not only are you allowing your body to move more fluidly with motion, but you're also not relying on one-half of your body.

Momentum is also a natural reaction of a player's Level of fluidity through impact in contrast to halting and/or stopping short. A player who is conditioned to move through their range and leverage the energy that has subsequently been produced, not only is the one with more control of their power but equally equipped to keep going longer due to having more stored energy. As such, momentum is one of many benefits of the role of functional movement patterns and your C.O.G and R.O.M are pivotal in the development of these key patterns of play. The 8 Keys embodies each of these principles and each respective key forms the next building block to progress towards a Top 10 ranking.

Week 4 of Block 1 Tasks

> **Episode** to listen to: #6
> **Book Chapters:**
> **Level 1:** *The Secrets to Optimal Performance Success (2016)* – 1.6
> **Level 2**: *What is Your Game Missing? (2019a)* – Ranking 4 *and* 4
> **Level 3**: *The Science of Elite Performance (2017)* – Chapter 1: Coach Communication

MATCHES/RESULTS

Opponent:	Opponent:	Opponent:
Tournament:	Tournament:	Tournament:
Score:	Score:	Score:
My Strength:	My Strength:	My Strength:
My Weakness:	My Weakness:	My Weakness:
Op Strength:	Op Strength:	Op Strength:
Op Weakness:	Op Weakness:	Op Weakness:
Other:	Other:	Other:

MATCHES/RESULTS

Opponent:	Opponent:	Opponent:
Tournament:	Tournament:	Tournament:
Score:	Score:	Score:
My Strength:	My Strength:	My Strength:
My Weakness:	My Weakness:	My Weakness:
Op Strength:	Op Strength:	Op Strength:
Op Weakness:	Op Weakness:	Op Weakness:
Other:	Other:	Other:

KEY TAKEAWAYS

Episode Notes:

Chapter Notes:

Topic Notes:

GOALS

What I would like to accomplish and/or improve this week:

What I am working on:

REFLECTION

Favourite part of the week I am most pumped for:

NOTES

TALK IT OUT

What I want to discuss with my coach this week (and note if I did discuss this and the outcome):

What I want to learn more about this week (and note if I did in fact learn more about this):

PROGRESS

Top 3 superpower moments:

Any goal updates?

Top 3 key learnings to bring forward:

Biggest motivator:

1

TRAINING SESSION DIARY

Monday:

Tuesday:

Wednesday:

Thursday:

Friday:

Saturday/Sunday:

IMPROVEMENTS:

What I want to improve?

Biggest distraction?

How was my nutrition?

MATCHES/RESULTS

Opponent:

Tournament:

Score:

My Strength:

My Weakness:

Op Strength:

Op Weakness:

Other:

Opponent:

Tournament:

Score:

My Strength:

My Weakness:

Op Strength:

Op Weakness:

Other:

Opponent:

Tournament:

Score:

My Strength:

My Weakness:

Op Strength:

Op Weakness:

Other:

TOURNAMENTS

Tournament I'm playing and/or following:

Favourite player to watch in-action:

NON-TENNIS

Highlights:

Hobby I am most enjoying:

NOTES

MY NUMBERS FOR THE WEEK

Hours on Court:

Hours on Fitness:

Work Ethic / 10:

My [Tennis] Black Book: Federer, Nadal, Serena, Djokovic, __________

KEY TAKEAWAYS

Episode Notes:

Chapter Notes:

Topic Notes:

GOALS

What I would like to accomplish and/or improve this week:

What I am working on:

TALK IT OUT

What I want to discuss with my coach this week (and note if I did discuss this and the outcome):

What I want to learn more about this week (and note if I did in fact learn more about this):

REFLECTION

Favourite part of the week I am most pumped for:

NOTES

PROGRESS

Top 3 superpower moments:

Any goal updates?

Top 3 key learnings to bring forward:

Biggest motivator:

2

TRAINING SESSION DIARY

Monday:

Tuesday:

Wednesday:

Thursday:

Friday:

Saturday/Sunday:

IMPROVEMENTS:

What I want to improve?

Biggest distraction?

How was my nutrition?

MATCHES/RESULTS

Opponent:

Tournament:

Score:

My Strength:

My Weakness:

Op Strength:

Op Weakness:

Other:

Opponent:

Tournament:

Score:

My Strength:

My Weakness:

Op Strength:

Op Weakness:

Other:

Opponent:

Tournament:

Score:

My Strength:

My Weakness:

Op Strength:

Op Weakness:

Other:

TOURNAMENTS

Tournament I'm playing and/or following:

Favourite player to watch in-action:

NON-TENNIS

Highlights:

Hobby I am most enjoying:

NOTES

MY NUMBERS FOR THE WEEK

Hours on Court:

Hours on Fitness:

Work Ethic / 10:

KEY TAKEAWAYS

Episode Notes:

Chapter Notes:

Topic Notes:

GOALS

What I would like to accomplish and/or improve this week:

What I am working on:

REFLECTION

Favourite part of the week I am most pumped for:

TALK IT OUT

What I want to discuss with my coach this week (and note if I did discuss this and the outcome):

What I want to learn more about this week (and note if I did in fact learn more about this):

NOTES

PROGRESS

Top 3 superpower moments:

Any goal updates?

Top 3 key learnings to bring forward:

Biggest motivator:

3

TRAINING SESSION DIARY

Monday:

Tuesday:

Wednesday:

Thursday:

Friday:

Saturday/Sunday:

TOURNAMENTS

Tournament I'm playing and/or following:

Favourite player to watch in-action:

IMPROVEMENTS:

What I want to improve?

Biggest distraction?

How was my nutrition?

MATCHES/RESULTS

Opponent:

Tournament:

Score:

My Strength:

My Weakness:

Op Strength:

Op Weakness:

Other:

Opponent:

Tournament:

Score:

My Strength:

My Weakness:

Op Strength:

Op Weakness:

Other:

Opponent:

Tournament:

Score:

My Strength:

My Weakness:

Op Strength:

Op Weakness:

Other:

NON-TENNIS

Highlights:

Hobby I am most enjoying:

NOTES

MY NUMBERS FOR THE WEEK

Hours on Court:

Hours on Fitness:

Work Ethic / 10:

My [Tennis] Black Book: Federer, Nadal, Serena, Djokovic, __________

KEY TAKEAWAYS

Episode Notes:

Chapter Notes:

Topic Notes:

GOALS

What I would like to accomplish and/or improve this week:

What I am working on:

REFLECTION

Favourite part of the week I am most pumped for:

TALK IT OUT

What I want to discuss with my coach this week (and note if I did discuss this and the outcome):

What I want to learn more about this week (and note if I did in fact learn more about this):

NOTES

PROGRESS

Top 3 superpower moments:

Any goal updates?

Top 3 key learnings to bring forward:

Biggest motivator:

4

TRAINING SESSION DIARY

Monday:

Tuesday:

Wednesday:

Thursday:

Friday:

Saturday/Sunday:

TOURNAMENTS

Tournament I'm playing and/or following:

Favourite player to watch in-action:

IMPROVEMENTS:

What I want to improve?

Biggest distraction?

How was my nutrition?

MATCHES/RESULTS

Opponent:

Tournament:

Score:

My Strength:

My Weakness:

Op Strength:

Op Weakness:

Other:

Opponent:

Tournament:

Score:

My Strength:

My Weakness:

Op Strength:

Op Weakness:

Other:

Opponent:

Tournament:

Score:

My Strength:

My Weakness:

Op Strength:

Op Weakness:

Other:

NON-TENNIS

Highlights:

Hobby I am most enjoying:

NOTES

MY NUMBERS FOR THE WEEK

Hours on Court:

Hours on Fitness:

Work Ethic / 10:

End of Month Block Review & Additional Notes

KEY TAKEAWAYS

GOALS

TALK IT OUT

REFLECTION

TRAINING SESSION DIARY

PROGRESS

NON-TENNIS

IMPROVEMENTS

NOTES

MATCHES/RESULTS

Opponent:
Tournament:
Score:
My Strength:
My Weakness:
Op Strength:
Op Weakness:
Other:

Opponent:
Tournament:
Score:
My Strength:
My Weakness:
Op Strength:
Op Weakness:
Other:

Opponent:
Tournament:
Score:
My Strength:
My Weakness:
Op Strength:
Op Weakness:
Other:

February: Block 2

GOALS:

Week 1 of Block 2 Tasks

> **Episodes** to listen to: #20 *and* #21
> **Book Chapters:**
> **Level 1:** *The Secrets to Optimal Performance Success (2016)* – 3.4 *and* 3.5
> **Level 2**: *What is Your Game Missing? (2019a)* – Ranking 5 *and* 6
> **Level 3**: *The Science of Elite Performance (2017)* – Chapter 1: Contact Zone / Ball Height

Topic: Functional Movement – Power and Balance

Now that you're familiar with *two* primary building blocks of functional movement patterns, additional offshoots are Power and Balance. The integration of functional movement patterns of play into your game has immense benefits. Not only the natural ability to help mitigate injuries but to *enhance* a player's natural power with near ease is courtesy of these respective patterns of play. The human body is comprised of something that is termed the kinetic chain, and this 'chain' is responsible for a player's capacity to perform a certain action and at what Level. Fundamental in its makeup, the kinetic chain is essentially a series of actions that take place if a certain movement triggers the said action. Think of it as a roll-on effect from one discrete skill to the next and how together multiple discrete skills form a serial skill. In this instance, by activating your C.O.G the human body naturally actives the lower body to engage both the top and bottom half i.e. the whole body. And in doing so, a player's R.O.M is then called on. Depending on the sequence of events, momentum will be a by-product and collectively, a player's ability to produce power is enhanced as is their ability to maintain a state of balance — in action, termed *dynamic* balance.

> These technical proponents work together and one without the other will weaken the 'chain' of events. By working on these technical proponents over a period of time to build a natural action-reaction response, a player's game will begin to move to the next Level.

This 'chain' of development is applicable at the elite Level of play through to the developmental player whilst the respective 'chain' of events will vary between players as will a player's capacity to produce power and maintain a certain Level of balance *before* being able to produce the desired outcome. Nevertheless, over time this will become more fine-tuned courtesy of a player's C.O.G and R.O.M progressing – as a more intrinsic action and/or reaction, and the adoption of momentum will become a collective by-product of power. Interestingly, neither momentum or power will become optimised without the ability to maintain a state of *dynamic* balance through play.

All these factors go into building an elite player and are synonymous with a player's Level of play inside the Top 10 when directly correlated *with* their performance. In other words, some players inside the Top 10 will *not* consistently correlate with this ranking range due to the disparity between their current Level of play and the 8%. As such, these players are a part of the 2% (or more) that will regress outside the Top 10 each season. To the contrary, when these factors are in play with the 8 Keys firmly integrated into a player's game, they become even more likely to not only further progress and solidify their place inside the 8%, but to achieve Grand Slam *wins* i.e. replicated success – *multiple* Grand Slams.

A Level of awareness of these principles is an absolute benchmark. The same applies for the 8 Keys and those who are not privy to these Keys are highly susceptible to regress in the rankings and scarcely progress beyond the Top 50. To break past this ranking threshold, several Keys need to be in place that will result in a player progressing towards the Top 30 and ultimately the Top 20. But not until *all* 7 Keys begin to formulate and become collectively activated will a player advance towards the Top 10. At this point, a player becomes reliant on the 8 Keys which are responsible for keeping a player inside the 8%. Intertwined within the 8 Keys and as a core foundation are these functional movement principles. Players who opt to *not* integrate and/or adopt these movement patterns will not only lag in performance but place themselves at a greater risk of *not* safeguarding their player-body against the onset of injury.

A bonus of integrating these functional movement patterns is the enhanced power and more versatile levels of dynamic balance that are ultimately at a player's disposal. This is a natural development of a player *learning* to leverage their kinetic chain to its full advantage. Whilst it doesn't happen overnight and is required to be consistently conditioned throughout The Long Game, by learning to integrate these core 'scientific' principles the player broadens their knowledge base and becomes even more capable of pushing their current Level of play to new and greater heights.

It's important to remember that whilst power typically is considered as a 'winning' component over balance, the Level of effectiveness to *maintenance* – consistency in application, is next to non-existent at the elite Level of play without a sturdy and well-rounded Level of (*dynamic*) balance. Simply put, one does not reach a new Level of optimisation without the other. Dynamic balance can be attributed to a player's rate of success in progressing towards a Top 10 ranking and coexists with a heightened Level of control of a player's C.O.G. Interestingly, a player's R.O.M called upon to maintain dynamic balance as flexion of the player-body is R.O.M in *action* which is needed for a player to lower their C.O.G to heighten their Level of control in *action*. As for power, it is essentially a by-product of the kinetic chain in *action* and the more 'parts' of the kinetic chain that are activated, the greater the possibility to produce power.

> Due to a common misconception around the production of power or rather the role of the kinetic chain often being overlooked, dynamic balance is not given as much attention as deserved and can be attributed in part to a lack of control around a player's C.O.G to R.O.M.

Without these two core functional movement principles, a player's risk of injury increases. This is due in part to the whole body not being activated per the kinetic chain and an unbalanced state then occurs with a greater load primarily placed on the player's upper body. As the lower body is responsible for lowering a player's C.O.G, a player who places the greater load on their upper body becomes 'top heavy' and is not positioned to optimise their kinetic chain. If continued and not modified, this can have detrimental long-term effects on the player-body not to mention a reduced Level of *optimised* performance.

Insofar as rankings are concerned, a player who ignores the 'power' of dynamic balance will more likely regress in the rankings and/or be halted from reaching their *next* peak performance in contrast to players who take advantage of its benefits. Putting in the time to build and establish a sound foundation of 'balance' will better equip a player in the long run during increased moments of pressure by preparing the player-body with the action-response to *stay* in play. Collectively, this type of balance directly feeds into a player's Level of control and in turn, a player's ability to *control* their power. Ultimately, balance remains fundamental in all functional movement proponents to optimise play. Not only does *dynamic* balance play a pivotal role in the underlying 8 Keys and is a central proponent of The V by Dr B (c), but it is also a definitive optimal performance metric to achieve a Top 10 ranking.

Week 2 of Block 2 Tasks

Episode to listen to: #10
Book Chapters:
Level 1: *The Secrets to Optimal Performance Success (2016)* – 2.3
Level 2: *What is Your Game Missing? (2019a)* – Ranking 6 *and* 7
Level 3: *The Science of Elite Performance (2017)* – Chapter 1: Groundstroke Technique

Week 3 of Block 2 Tasks

Episodes to listen to: #22 *and* #26
Book Chapters:
Level 1: *The Secrets to Optimal Performance Success (2016)* – 4.1 *and* 4.5
Level 2: *What is Your Game Missing? (2019a)* – Ranking 7 *and* 8
Level 3: *The Science of Elite Performance (2017)* – Chapter 2: Athlete Anticipation and the Role of Perception in Performing the Quintessential Technique

Topic: Advantages, Engagement and Re-learning Skills

An often-overlooked concept is the dynamic that our *experience* plays in forming our own biases in respect to our sporting background. From a tennis player in a singular-player context to a basketball player in a multi-player environment, both serve as preliminaries for what's to come. But this intermediary relationship between sports and a player's involvement – individual to team interactions and the value on offer when these become a staple in a player's development, is often forgotten.

Irrespective of the sport, there is a skill to be learned. Through this learning a player becomes more conscious of the 'skills' on offer and the potential tools at their disposal. By simply being exposed to a multitude of sports in varying environments – singular to multi, a player will inadvertently 'pocket' some of these new learnings for recall in their later development. And this is a continued cycle for players at varying levels. The player who is not shy of experiencing new sporting contexts will equally be open to consider *how* and/or *if* these new 'learnings' can be integrated in their game over time.

But it doesn't always need to be a tool that is game-based and/or tactically/strategically centred for a player to benefit. For the singular player, they are fortunate that a quick transition to doubles play is readily accessible but this is still a relatively small dynamic with limited cross-sport applications (given they are still playing the *same* sport in a different format). By changing the environment from a tennis court to a basketball court, baseball field or to countless other sport-centred environments, the player is forced to learn 'new' skills that are not essentially aligned with their 'expertise'. Sure, the ability to play the sport may be present, but what you're looking for here is the ability to learn team-based skills – the way a baseball team *sees* the field, how a basketball team prepares for their opponent, to how a soccer team *reads* the next play. There will undoubtedly be crossovers and an evident knowledge-share but on occasion, there will be key take-homes you may very well be able to apply in your own game. As an added advantage, a player will increase their overall skill set by simply participating in more than one sport to increase their conditioning whilst not being inside their tennis-centred comfort zone.

> This serves a purpose when a player is under pressure in match-based scenarios and pushed outside their comfort zone with the indirect conditioned ability to still function at a high Level when under pressure.

For the team-based athlete, they are often reliant on the support from their teammates and are conditioned to work 'with' their team to achieve a desired outcome. For the singular athlete they are conditioned to rely on themselves and do not have a 'team' on court at the same time to help them achieve their set objective – it is just *you* and no one else whilst acknowledging that whilst on-court coaching does happen, is *not* a universal practice at this time.

You may think that since tennis is a singular sport, learning from a team-based sport may not be advantageous. But learning from multiple players on the court and/or field at the same time and learning *how* this knowledge-share takes place to *how* one athlete performs quite differently to another and *why*, there are significant crossovers here to consider for the player to apply in their own game – now and/or in the future.

Learning in a multi-player environment for a single-sport athlete challenges a player's skill acquisition to verbal and non-verbal communication skills. The opposite is the case for the multi-player athlete placed into a singular environment. A tennis player can leverage various sports whilst forming and conditioning new skills, periodically, that have the capacity both directly and indirectly to optimise their performance toolkit. And this is what separates the best players in the world.

New skills can be attained across various sports, whether in respect to enhancing a player's current performance and/or overall motor skills. There are also other benefits overlooked when a player engages in various sports that range from perception, anticipation to new methods of thinking to achieve a desired outcome. One skill that is universal across sports is problem solving. In tennis, the player will have become readily accustomed to solving on-court problems and as such, this skill to 'solve-on-demand' is highly attuned. When a player is placed in an opposing environment, this ability to solve is not as readily available as the player needs to learn – quickly, *how* to achieve the desired outcome with new 'pressure' to deal with and manage. This experience in a non-tennis context engages the player in a non-typical way and inadvertently fosters the ability to solve problems on and off the court. This is one of many examples of how an athlete's Level of engagement is further enhanced when the ideology behind single sport interactions embraces the multi-player environment and serves as a new training metric.

Have you ever wondered why players who embark on achieving Championship success in the Junior ranks scarcely achieve the same results on the WTA and/or ATP Tours? If the same Pathway is followed from the Junior ranks to the Senior ranks, the Grand Slam results would arguably be the same, however, this is seldom the case. Players who are fortunate enough to 'win' Junior Grand Slam Championships are rarely the ones to achieve this status once they've joined their respective Tour. The irony here is that those who do not peak as soon or achieve the pinnacle of success in their developmental years, these players often are the ones who progress more steadily opposed to being 'stuck' at a certain Level and inside a specific ranking range i.e. a performance plateau inside the Top 100 but outside the Top 80 for several seasons.

When a player in their developmental years peaks too soon with rapid fire results, this comes with its own set of risks. In respect to the collective development of a player's game, there will be discrete skills that need to continue to be progressed to compound and further optimise a specific serial skill. Unfortunately, players who have these rapid-fire results typically neglect to integrate fundamental discrete skills and their player-body compensates for this oversight by compromising their fundamental mechanics to achieve the desired outcome.

It can be frustrating for a player who achieves heightened success in the junior ranks to then not achieve a similar Level of success in the senior ranks. For newer players who have emerged inside the Top 100 on the WTA or ATP Tours, there is a reason why only a set number of these players will continue to ascend towards the Top 50 inside their initial two seasons. By all accounts, some players may progress inside their initial season but there is caution to be taken as if the same rapid-fire results are achieved and the same mistakes made, they remain just as susceptible to a steep regression due to their body being overloaded and longevity at the top simply not being sustainable until their foundational mechanics have been re-learned and restored.

But there's a surefire way to avoid these associated risks. First and foremost, the 7 Keys progressively need to be integrated into a player's game and if these are ignored, a ranking plateau outside the Top 80 will soon follow. Similarly, for the junior with rapid-fire results that achieves a Top 100 ranking and does not commit to restore and/or re-learn new technical parameters to take their game to the next Level, these players will seldom breach the Top 50. Unfortunately, the ideology behind learning and the need to re-learn a skill – discrete and/or serial, is often overlooked not as a means of a progression but seen as a regression. What becomes ignored here is that the need to re-learn and/or restore is a continuation of one discrete skill to the next. For example, integrating a lowered C.O.G whilst keeping all other elements of a player's game intact takes a lot of work. Then to increase the use of 'work' a player can apply in one serial skill takes a lot of effort.

> Just these two aspects of a player's game may very well take 4-6 weeks to integrate and potentially a further 4-6 months to reach a Level of autonomy and form a new baseline. Typically, it will take much longer and this is an example of simple skill integrations for the elite player inside the Top 30.

When a developmental player is pushed too much too soon, their capacity to round out their game is not of priority and is overlooked. But this needs to be done sooner opposed to later *if* that player aspires to progress towards the Top 50 let alone the Top 10 in their initial seasons on Tour. Whilst any player can progress later in their career towards a Top 10 ranking, this is specifically in reference to the junior player and their ability to restore and re-learn movement patterns. Through the developmental period of these junior players, they've often set themselves up for failure in their later years – those initial seasons on Tour, and to succumb to injury due to the pre-conditioned movements in their earlier years *not* being restored – modified and re-learned.

This 'hidden' injury typically prevents a player from progressing towards that next Level of play as their current technical parameters do not have the capacity to achieve the desired results. To overcome this mishap, the 7 Keys are absolute and the 8 Keys will reinforce their rulers to place a player back on the Pathway towards a Top 10 ranking. By affording a player a better and more informed knowledge source around the conditioning process, this will ensure *you* are more prepared for this restoration stage and why the role of re-learning is a positive step.

Whilst the transition from junior to senior player often leaves players more vulnerable and caught unaware of the 'poor' technical foundations that have been rapidly built to deliver rapid-fire results, the *work* now needs to be done to protect the player – allowing you to have a long and fulfilling career opposed to spending your initial years on Tour sidelined. To ensure the 'work' is done, take caution in knowing the right principles put into place early are best, but if the damage has been done, restoring your true movement patterns whilst continuously working on your discrete and/or serial skills remains key. In time, the 8 Keys will become embedded in your game – should you follow the rulers of the *Collection*, and optimal performance successes will follow.

Week 4 of Block 2 Tasks

> **Episode** to listen to: #11
> **Book Chapters:**
> **Level 1:** *The Secrets to Optimal Performance Success (2016)* – 2.4
> **Level 2**: *What is Your Game Missing? (2019a)* – Ranking 8 *and* 9
> **Level 3**: *The Science of Elite Performance (2017)* – Chapter 2: Centre of Gravity and its Association with a Players Contact Zone during the Technique

MATCHES/RESULTS

Opponent:	Opponent:	Opponent:
Tournament:	Tournament:	Tournament:
Score:	Score:	Score:
My Strength:	My Strength:	My Strength:
My Weakness:	My Weakness:	My Weakness:
Op Strength:	Op Strength:	Op Strength:
Op Weakness:	Op Weakness:	Op Weakness:
Other:	Other:	Other:

MATCHES/RESULTS

Opponent:	Opponent:	Opponent:
Tournament:	Tournament:	Tournament:
Score:	Score:	Score:
My Strength:	My Strength:	My Strength:
My Weakness:	My Weakness:	My Weakness:
Op Strength:	Op Strength:	Op Strength:
Op Weakness:	Op Weakness:	Op Weakness:
Other:	Other:	Other:

MATCHES/RESULTS

Opponent:	Opponent:	Opponent:
Tournament:	Tournament:	Tournament:
Score:	Score:	Score:
My Strength:	My Strength:	My Strength:
My Weakness:	My Weakness:	My Weakness:
Op Strength:	Op Strength:	Op Strength:
Op Weakness:	Op Weakness:	Op Weakness:
Other:	Other:	Other:

My [Tennis] Black Book: Federer, Nadal, Serena, Djokovic, __________

KEY TAKEAWAYS

Episode Notes:

Chapter Notes:

Topic Notes:

GOALS

What I would like to accomplish and/or improve this week:

What I am working on:

REFLECTION

Favourite part of the week I am most pumped for:

TALK IT OUT

What I want to discuss with my coach this week (and note if I did discuss this and the outcome):

What I want to learn more about this week (and note if I did in fact learn more about this):

NOTES

PROGRESS

Top 3 superpower moments:

Any goal updates?

Top 3 key learnings to bring forward:

Biggest motivator:

1

TRAINING SESSION DIARY

Monday:

Tuesday:

Wednesday:

Thursday:

Friday:

Saturday/Sunday:

TOURNAMENTS

Tournament I'm playing and/or following:

Favourite player to watch in-action:

IMPROVEMENTS:

What I want to improve?

Biggest distraction?

How was my nutrition?

MATCHES/RESULTS

Opponent:

Tournament:

Score:

My Strength:

My Weakness:

Op Strength:

Op Weakness:

Other:

Opponent:

Tournament:

Score:

My Strength:

My Weakness:

Op Strength:

Op Weakness:

Other:

Opponent:

Tournament:

Score:

My Strength:

My Weakness:

Op Strength:

Op Weakness:

Other:

NON-TENNIS

Highlights:

Hobby I am most enjoying:

NOTES

MY NUMBERS FOR THE WEEK

Hours on Court:

Hours on Fitness:

Work Ethic / 10:

KEY TAKEAWAYS

Episode Notes:

Chapter Notes:

Topic Notes:

GOALS

What I would like to accomplish and/or improve this week:

What I am working on:

REFLECTION

Favourite part of the week I am most pumped for:

TALK IT OUT

What I want to discuss with my coach this week (and note if I did discuss this and the outcome):

What I want to learn more about this week (and note if I did in fact learn more about this):

NOTES

Top 3 superpower moments:

Any goal updates?

PROGRESS

TRAINING SESSION DIARY

Monday:

Top 3 key learnings to bring forward:

Biggest motivator:

2

TOURNAMENTS

Tournament I'm playing and/or following:

Tuesday:

IMPROVEMENTS:

What I want to improve?

MATCHES/RESULTS

Opponent:

Tournament:

Score:

My Strength:

My Weakness:

Op Strength:

Op Weakness:

Other:

Wednesday:

Favourite player to watch in-action:

Biggest distraction?

Thursday:

NON-TENNIS

Highlights:

How was my nutrition?

Friday:

Hobby I am most enjoying:

Saturday/Sunday:

Opponent:

Tournament:

Score:

My Strength:

My Weakness:

Op Strength:

Op Weakness:

Other:

Opponent:

Tournament:

Score:

My Strength:

My Weakness:

Op Strength:

Op Weakness:

Other:

NOTES

MY NUMBERS FOR THE WEEK

Hours on Court:

Hours on Fitness:

Work Ethic / 10:

KEY TAKEAWAYS

Episode Notes:

Chapter Notes:

Topic Notes:

GOALS

What I would like to accomplish and/or improve this week:

What I am working on:

REFLECTION

Favourite part of the week I am most pumped for:

TALK IT OUT

What I want to discuss with my coach this week (and note if I did discuss this and the outcome):

What I want to learn more about this week (and note if I did in fact learn more about this):

NOTES

PROGRESS

Top 3 superpower moments:

Any goal updates?

Top 3 key learnings to bring forward:

Biggest motivator:

3

TRAINING SESSION DIARY

Monday:

Tuesday:

Wednesday:

Thursday:

Friday:

Saturday/Sunday:

TOURNAMENTS

Tournament I'm playing and/or following:

Favourite player to watch in-action:

IMPROVEMENTS:

What I want to improve?

Biggest distraction?

How was my nutrition?

MATCHES/RESULTS

Opponent: ____________

Tournament: ____________

Score: ____________

My Strength: ____________

My Weakness: ____________

Op Strength: ____________

Op Weakness: ____________

Other: ____________

Opponent: ____________

Tournament: ____________

Score: ____________

My Strength: ____________

My Weakness: ____________

Op Strength: ____________

Op Weakness: ____________

Other: ____________

Opponent: ____________

Tournament: ____________

Score: ____________

My Strength: ____________

My Weakness: ____________

Op Strength: ____________

Op Weakness: ____________

Other: ____________

NON-TENNIS

Highlights:

Hobby I am most enjoying:

NOTES

MY NUMBERS FOR THE WEEK

Hours on Court:

Hours on Fitness:

Work Ethic / 10:

KEY TAKEAWAYS

Episode Notes:

Chapter Notes:

Topic Notes:

GOALS

What I would like to accomplish and/or improve this week:

What I am working on:

TALK IT OUT

What I want to discuss with my coach this week (and note if I did discuss this and the outcome):

What I want to learn more about this week (and note if I did in fact learn more about this):

REFLECTION

Favourite part of the week I am most pumped for:

PROGRESS

Top 3 superpower moments:

Any goal updates?

Top 3 key learnings to bring forward:

Biggest motivator:

4

NOTES

TRAINING SESSION DIARY

Monday:

Tuesday:

Wednesday:

Thursday:

Friday:

Saturday/Sunday:

MY NUMBERS FOR THE WEEK

Hours on Court:

Hours on Fitness:

Work Ethic / 10:

IMPROVEMENTS:

What I want to improve?

Biggest distraction?

How was my nutrition?

MATCHES/RESULTS

Opponent:

Tournament:

Score:

My Strength:

My Weakness:

Op Strength:

Op Weakness:

Other:

Opponent:

Tournament:

Score:

My Strength:

My Weakness:

Op Strength:

Op Weakness:

Other:

Opponent:

Tournament:

Score:

My Strength:

My Weakness:

Op Strength:

Op Weakness:

Other:

TOURNAMENTS

Tournament I'm playing and/or following:

Favourite player to watch in-action:

NON-TENNIS

Highlights:

Hobby I am most enjoying:

NOTES

End of Month Block Review & Additional Notes

KEY TAKEAWAYS | GOALS | TRAINING SESSION DIARY | PROGRESS

TALK IT OUT | REFLECTION | NON-TENNIS | IMPROVEMENTS

NOTES

MATCHES/RESULTS

Opponent:
Tournament:
Score:
My Strength:
My Weakness:
Op Strength:
Op Weakness:
Other:

Opponent:
Tournament:
Score:
My Strength:
My Weakness:
Op Strength:
Op Weakness:
Other:

Opponent:
Tournament:
Score:
My Strength:
My Weakness:
Op Strength:
Op Weakness:
Other:

March: Block 3

GOALS:

Week 1 of Block 3 Tasks

Episodes to listen to: #25 *and* #27
Book Chapters:
Level 1: *The Secrets to Optimal Performance Success (2016)* – 4.4 *and* 4.6
Level 2: *What is Your Game Missing? (2019a)* – Ranking 9 *and* 10
Level 3: *The Science of Elite Performance (2017)* – Chapter 2: The Coach-Athlete Relationship and the Information Process

Topic: The Role of Education

A topic that is not given the attention it deserves, education can play an incredibly powerful role in a player's development. This can come in varying forms — from 'whatever' works for the player and draws a positive response, to integrating new methods of learning on various topics that will either have a direct or indirect impact on the player's knowledge bank. This 'knowledge bank' has been termed in earlier releases and features within the broader *Collection* as an absolute pivotal guidepost to expand a player's ability to perform – on and/or off the court with direct optimal performance correlations.

To help a player leverage their education it begins with the right guidance to allow this 'balancing act' to flow through to their performance. This corresponds with the Pathway and what's in place specifically for that player on the road ahead. More importantly, the Pathway was designed to help *you* – to guide you when perhaps that guidance either is not readily available and/or you're still in the process of acquiring the 8 Keys. Either way, it is pivotal for a player to know where to turn and where that guidance is coming from and this goes hand-in-hand with education.

One such form of education in respect to a tennis player's trajectory – whether to achieve a ranking inside the Top 500 or Top 50, is *planning* and is embedded in *your* Black Book. Right now, between your palms is an immense toolkit that will take you from one season to the next – in full. **My [Tennis] Black Book** is a staple of the broader *Collection* and it is that very *Collection* and its educational underpinning that will inform your knowledge bank and broader scope of learning – what it really takes to ascend towards a Top 10 ranking with the respective intervals between. It is of no surprise then that each Book that has been before **My [Tennis] Black Book** serves as its own purpose in guiding you from one *stage* of development to the next. And this applies to all levels of play – data that may be completely new and/or key learnings that are reinforced.

But there's so much more. External learning – education in the standardised context, has significant rewards that should be encouraged. The player who continues their studies, irrespective of the field, is continuously activating their ability to *learn* and whether directly or indirectly, solve problems. This constancy of adding to your scope of learning pays dividends through on court play. The player with this scope of learning has a broader knowledge bank to call upon when in need, whether that refers to point construction, understanding the parameters of your own movement and how you can consider the next Level of optimisation, to pinpointing your opponents' limits and how you can better construct your game *plan* to take further advantage.

Education allows a player to offset their performance in a way that draws on their versatile learnings — from problem solving skills (mathematical to science-based subjects) to creative skills (humanities to artistic subjects). Of course, these are mere examples and there are ample subjects to choose from but what is important here is that your game in fact *can* and *will* benefit from heightened levels of knowledge. It is also important to be mindful of the nature of various subjects and in this case, there is no 'must' – there is no current correlation between heightened learning in one subject that transfers to an optimised performance, asides from perhaps physical education. Adding to your *education* and in turn your knowledge bank through various subject areas will expand your scope and depth of understanding. Over time your knowledge bank will deepen, and you'll soon see how one subject serves the next and can form a solution for an entirely different field. And that is the *power* of education at all levels of play.

> Excitingly, if this trajectory continues and coincides with your ranking's trajectory, a player with this depth and scope inside the Top 10 has a clear ruler in common – *replicated success*.

Conversely, the developmental player can use education and their tennis performance to inadvertently help recall in a specific subject area. The same applies for the elite player and an area of study they are continuing and/or have chosen to add to their knowledge bank. By using points of recall in game-based situations, education is reciprocal to performance, and both can go hand-in-hand if desired. Through encouraging a player to 'think outside the box' whilst using their sport of choice with another dimension of learning, this can serve to compound and further optimise a player's current playing capacity.

The interrelationship between education and play can also bridge the gap often seen between students in their later formative years and a Level of disengagement in one or more subjects. Although this is generalised to accommodate standardised benchmarks, typically there is always a subject (or two) that is not as compelling as others. By using these subjects in game-based situations there is a potential opportunity for increased engagement and ultimately to add to the player's collective education and success in that given subject. By no means limited to one subject or two, if a player looks to increase engagement rates and *results*, game-based performances can enhance results both on and off the court.

The purpose of integrating education into *play* is to enhance a player's performance and their levels of engagement. If signs of disengagement appear, you know you've got too far – as with all components, the *balancing* act remains true. A player should always remain in control of their environment and subject of choice. The balance between engaging optimal performance outcomes to disengaging the player should be taken seriously to ensure ample benefit is delivered opposed to compromised. Ultimately, by building alternative levels of engagement from an educational perspective, this has the capacity to take the player to the 'next' Level as it signals a Level of willingness and preparedness to embrace the learnings around the 8 Keys and to move their current peak performance baseline to that *next* stage of optimisation.

Week 2 of Block 3 Tasks

> **Episode** to listen to: #12
> **Book Chapters:**
> **Level 1:** *The Secrets to Optimal Performance Success (2016)* – 2.5
> **Level 2**: *What is Your Game Missing? (2019a)* – Ranking 11 *and* 12
> **Level 3**: *The Science of Elite Performance (2017)* – Chapter 2: Optimal Contact Zone Required to Hit a Forehand

Week 3 of Block 3 Tasks

Episodes to listen to: #29 *and* #31
Book Chapters:
Level 1: *The Secrets to Optimal Performance Success (2016)* – 4.8 *and* 4.10
Level 2: *What is Your Game Missing? (2019a)* – Ranking 12 *and* 13
Level 3: *The Science of Elite Performance (2017)* – Chapter 2: Knee Mechanics and Influence in Performing the Technique

Topic: What's Your Purpose?

A question either asked or pondered by every tennis player when they arrive at a training session and/or step onto a court revolves around what's your '*why*' – here and now. Asides from the obvious of a more simplistic fun and/or enjoyment perspective, what is really being asked is *"what do I need to do now to edge me closer towards a Top 10 ranking"* – be it this season or next decade. Irrespective of the timeframe, every player should be asking themselves "What's *my* Purpose" every training session to ensure the work that needs to be done is at the forefront as *your* guidepost. Whilst players typically trust their coaches to be responsible for their *'why'* this lacks control and places your timeline at your coach's discretion and is not conducive of an optimal coach-athlete relationship.

From a philosophical perspective this question can apply to various aspects of life and/or life itself – but that's not the premise of **My [Tennis] Black Book**. Rather, the most important questions are based around *'what'* should I be doing *'now'* to edge me closer towards that Top 10 ranking? To answer that is the essence of *your* Black Book – to prompt you continuously throughout the duration of an entire season. But you are not expected to have an answer *all* the time. Some days will require more thought than others but if you follow *each plan each* week the answer will be before you – guaranteed. The time for concern will be when the same question is nor asked nor given the attention it deserves.

Asking is key and yet if you're not prepared to put time aside to provide an answer, then the question is merely theoretical. Explicitly in regard to what's before you when it comes to stepping *onto* the court – opposed to a deep dive that has your mind falling down a (metaphorical) rabbit hole, 'What's *my* Purpose' is a simple and yet powerful personal question. In the context of a tennis court, it becomes simpler as it follows suit with *your* Black Book and the Pathway. In order to break that down a step further, ask yourself *what* you are working on today and *why* then pen down your thoughts. *What* do you want to work on today and *why* follows – add to those thoughts. Do you agree with what your coach wants you to work on – another answer to pen down here or in the adjoining *Journal*. Why or why not – continue writing. Does your coach want to work on what you want to work on? Why or why not – that's quite a bit now to add to your writings *this* week and it is at your discretion whether you'll continue to add this detail each week or each Block.

Of course, there are more questions you can ask yourself around *your* purpose, including whether you are having an open discussion with your coach about what you're working on today – why and/or why not? If you're not working with a coach for today's session, what have they encouraged you to work on and *why* or if you're not currently working with a coach, what *should* you be working on today if you've been following the *Collection*? These questions can be framed in whatever light works best for you – they are personal to your *stage* and performance but maintain the same underlying intent – that's the purpose.

One question that is seldom asked and yet incredibly important is the question/s you ask yourself when no one else is around. This is in reference to *your* purpose – *your* intent when you step onto the court. Whether by yourself and/or with a hitting partner or coach, what do you think you should be doing and *why*? Be careful that you're not merely following blind instruction – really *know* your answer and the underlying why. Most players will fall into this 'trap' and before they know it, they are become reliant on their coach rather than maintaining that *balance* and their own personal power in the coach-athlete relationship.

> A Level of reliance built on 'not' knowing will never allow you to progress your game-based knowledge and to know right from wrong. Whilst there are universal rulers that govern the game of tennis, there is no one ideology that coaches adhere to and abide by (this is something the *Collection* addresses to empower your own performance).

Until you know that what is being shared coincides with and/or feeds into the learning and integration of the 8 Keys into your game, and follows the Pathway, it remains your responsibility to proactively *learn*. Without putting in the work to know what is necessary and what actionable steps to take to ensure you stay on track towards reaching one peak performance after the next, you remain susceptible to being blindsided and steered off track or to potentially develop an injury due to the rulers of the 8 Keys *not* being followed. But you're in the right place and ahead of the masses courtesy of **My [Tennis] Black Book** – the intention is to ensure you stick to the *plans* and remain wary of a Level of complacency.

The role of 'purpose' itself is often overlooked when it comes to the tennis player let alone asked of a player – have *you* been asked? Whilst this may come as a surprise, if a player is *not* asked what their purpose is and provided an open dialogue to discuss this and share as they choose, the communication between player and coach has already broken down before a coach-athlete relationship could even be formed – a significant red flag. A coaches role includes guiding the player towards their *tennis* purpose. In contrast, the purpose behind a player's actions in *life* may vary to their actual behaviour in training. And that's the point. A purpose is unique as it is to the person as it is to their environment. For a coach to genuinely engage with and understand their player – for *you* to be understood, requires your *why* to be shared as you step onto the court.

If you're slightly stumped, that's okay. The idea of conditioning 'purpose' – to build a habit to *ask* yourself *your* underlying *why*, fosters a greater Level of on-court awareness and ultimately places you in control of what next step is right for you. Of course, you're not expected to have all the answers – that's what the *Collection* is for and why *your* Black Book will prompt you to share your purpose repeatedly across the season. It is fundamental to ask yourself, in whatever shape it may feel natural for you, each time you step foot onto the court – be it "What's my intent?", "What's my Purpose?", or simply "What am I working on today and why?"

> A form of behavioural conditioning conducive to a player's performance, these questions also empower the player to know their underlying why and to take control.

This Level of control will give you over time – dependent on your current Level, a better understanding of the multifaceted demands in place and what is needed to head towards a Top 10 ranking. For example, whilst you may be familiar with the 8 Keys, you may *not* know how to embed each key into your game. This Level of awareness gives you the tools to know who you need to help you further develop your game i.e. a coach that is familiar with the 8 Keys and one that understand the importance of putting in the time to build a sound coach-athlete relationship. Whilst not every coach is for everyone, there is a recipe to build a successful partnership and both coach and player must put in the work. If you do not feel as though you're being heard, let alone asked about *your* purpose – you have your answer. But if the willingness is there on both sides and the 8 Keys are acquired and integrated, you're heading in the right direction – if not, your ranking is set to plateau outside the Top 80 irrespective of which Tour.

Unfortunately, it is largely common practice for a player to *not* have an answer to their 'purpose' let alone be prompted for one before, during and/or after their training. If this is you, you now have the tools to know better and the steps to take – use **My [Tennis] Black Book** as your guide. It's fundamental to remain conscious of the 8 Keys and their solidification into your game on a cyclical basis i.e. there's no 'off' and they should remain as a constant in your game to reach the next 'new' Level of autonomy. Factoring *a* key into your purpose and/or a part of your purpose on a rolling basis is a good place to start when and/or if in doubt of your purpose, today, tomorrow or this week. But be sure to personalise your purpose for *you*. By devising a purpose, you can better prepare yourself to achieve not merely your performance goals but equally have a constant guidepost to reference whilst proactively practicing with intent to develop your next optimal performance outcome and achieve that *next* performance milestone.

Week 4 of Block 3 Tasks

> **Episode** to listen to: #14
> **Book Chapters:**
> **Level 1:** *The Secrets to Optimal Performance Success (2016)* – 2.7
> **Level 2**: *What is Your Game Missing? (2019a)* – Ranking 14 *and* 15
> **Level 3**: *The Science of Elite Performance (2017)* – Chapter 3: Effective Communication with an Athlete when Learning a New Skill

KEY TAKEAWAYS

Episode Notes:

Chapter Notes:

Topic Notes:

GOALS

What I would like to accomplish and/or improve this week:

What I am working on:

REFLECTION

Favourite part of the week I am most pumped for:

TALK IT OUT

What I want to discuss with my coach this week (and note if I did discuss this and the outcome):

What I want to learn more about this week (and note if I did in fact learn more about this):

NOTES

PROGRESS

Top 3 superpower moments:

Any goal updates?

Top 3 key learnings to bring forward:

Biggest motivator:

1

TRAINING SESSION DIARY

Monday:

Tuesday:

Wednesday:

Thursday:

Friday:

Saturday/Sunday:

IMPROVEMENTS:

What I want to improve?

Biggest distraction?

How was my nutrition?

MATCHES/RESULTS

Opponent:

Tournament:

Score:

My Strength:

My Weakness:

Op Strength:

Op Weakness:

Other:

Opponent:

Tournament:

Score:

My Strength:

My Weakness:

Op Strength:

Op Weakness:

Other:

Opponent:

Tournament:

Score:

My Strength:

My Weakness:

Op Strength:

Op Weakness:

Other:

TOURNAMENTS

Tournament I'm playing and/or following:

Favourite player to watch in-action:

NON-TENNIS

Highlights:

Hobby I am most enjoying:

NOTES

MY NUMBERS FOR THE WEEK

Hours on Court:

Hours on Fitness:

Work Ethic / 10:

KEY TAKEAWAYS

Episode Notes:

Chapter Notes:

Topic Notes:

GOALS

What I would like to accomplish and/or improve this week:

What I am working on:

REFLECTION

Favourite part of the week I am most pumped for:

TALK IT OUT

What I want to discuss with my coach this week (and note if I did discuss this and the outcome):

What I want to learn more about this week (and note if I did in fact learn more about this):

NOTES

Top 3 superpower moments:

Any goal updates?

PROGRESS

TRAINING SESSION DIARY

Monday:

Tuesday:

Wednesday:

Thursday:

Friday:

Saturday/Sunday:

Top 3 key learnings to bring forward:

Biggest motivator:

2

TOURNAMENTS

Tournament I'm playing and/or following:

Favourite player to watch in-action:

IMPROVEMENTS:

What I want to improve?

Biggest distraction?

How was my nutrition?

MATCHES/RESULTS

Opponent:

Tournament:

Score:

My Strength:

My Weakness:

Op Strength:

Op Weakness:

Other:

NON-TENNIS

Highlights:

Hobby I am most enjoying:

Opponent:

Tournament:

Score:

My Strength:

My Weakness:

Op Strength:

Op Weakness:

Other:

Opponent:

Tournament:

Score:

My Strength:

My Weakness:

Op Strength:

Op Weakness:

Other:

NOTES

MY NUMBERS FOR THE WEEK

Hours on Court:

Hours on Fitness:

Work Ethic / 10:

KEY TAKEAWAYS

Episode Notes:

Chapter Notes:

Topic Notes:

GOALS

What I would like to accomplish and/or improve this week:

What I am working on:

TALK IT OUT

What I want to discuss with my coach this week (and note if I did discuss this and the outcome):

What I want to learn more about this week (and note if I did in fact learn more about this):

REFLECTION

Favourite part of the week I am most pumped for:

PROGRESS

Top 3 superpower moments:

Any goal updates?

Top 3 key learnings to bring forward:

Biggest motivator:

3

NOTES

TRAINING SESSION DIARY

Monday:

Tuesday:

Wednesday:

Thursday:

Friday:

Saturday/Sunday:

IMPROVEMENTS:

What I want to improve?

Biggest distraction?

How was my nutrition?

MATCHES/RESULTS

Opponent:

Tournament:

Score:

My Strength:

My Weakness:

Op Strength:

Op Weakness:

Other:

Opponent:

Tournament:

Score:

My Strength:

My Weakness:

Op Strength:

Op Weakness:

Other:

Opponent:

Tournament:

Score:

My Strength:

My Weakness:

Op Strength:

Op Weakness:

Other:

TOURNAMENTS

Tournament I'm playing and/or following:

Favourite player to watch in-action:

NON-TENNIS

Highlights:

Hobby I am most enjoying:

NOTES

MY NUMBERS FOR THE WEEK

Hours on Court:

Hours on Fitness:

Work Ethic / 10:

KEY TAKEAWAYS

Episode Notes:

Chapter Notes:

Topic Notes:

GOALS

What I would like to accomplish and/or improve this week:

What I am working on:

REFLECTION

Favourite part of the week I am most pumped for:

NOTES

TALK IT OUT

What I want to discuss with my coach this week (and note if I did discuss this and the outcome):

What I want to learn more about this week (and note if I did in fact learn more about this):

PROGRESS

Top 3 superpower moments:

Any goal updates?

Top 3 key learnings to bring forward:

Biggest motivator:

4

TRAINING SESSION DIARY

Monday:

Tuesday:

Wednesday:

Thursday:

Friday:

Saturday/Sunday:

TOURNAMENTS

Tournament I'm playing and/or following:

Favourite player to watch in-action:

IMPROVEMENTS:

What I want to improve?

Biggest distraction?

How was my nutrition?

MATCHES/RESULTS

Opponent: ____________

Tournament: ____________

Score: ____________

My Strength: ____________

My Weakness: ____________

Op Strength: ____________

Op Weakness: ____________

Other: ____________

Opponent: ____________

Tournament: ____________

Score: ____________

My Strength: ____________

My Weakness: ____________

Op Strength: ____________

Op Weakness: ____________

Other: ____________

Opponent: ____________

Tournament: ____________

Score: ____________

My Strength: ____________

My Weakness: ____________

Op Strength: ____________

Op Weakness: ____________

Other: ____________

NON-TENNIS

Highlights:

Hobby I am most enjoying:

NOTES

MY NUMBERS FOR THE WEEK

Hours on Court:

Hours on Fitness:

Work Ethic / 10:

End of Month Block Review & Additional Notes

KEY TAKEAWAYS

GOALS

TALK IT OUT

REFLECTION

TRAINING SESSION DIARY

PROGRESS

NON-TENNIS

IMPROVEMENTS

NOTES

MATCHES/RESULTS

Opponent:
Tournament:
Score:
My Strength:
My Weakness:
Op Strength:
Op Weakness:
Other:

Opponent:
Tournament:
Score:
My Strength:
My Weakness:
Op Strength:
Op Weakness:
Other:

Opponent:
Tournament:
Score:
My Strength:
My Weakness:
Op Strength:
Op Weakness:
Other:

April: Block 4

GOALS:

Week 1 of Block 4 Tasks

> **Episodes** to listen to: #30 *and* #39
> **Book Chapters:**
> **Level 1:** *The Secrets to Optimal Performance Success (2016)* – 4.9 *and* 6.1
> **Level 2**: *What is Your Game Missing? (2019a)* – Ranking 15 *and* 16
> **Level 3**: *The Science of Elite Performance (2017)* – Chapter 3: Coach Behaviour and Athlete Responsiveness

Topic: Learning from Mistakes

Everyone makes mistakes. It's one of those things in life that is inevitable. How you respond to those mistakes will vary and these behaviours are often formed in your developmental years. Only later in life, if and/or when you choose, will you begin to modify these behaviours. On this premise, it is so incredibly important to set healthy precedents early to ensure your response is well-attuned and builds the 'right' kind of behaviours opposed to setting yourself up for more work in the years ahead i.e. needing to 'rewire' these behaviours due to poorly developed responses. This is simple human behaviour which none of us are immune from and the conditioning response for good and/or bad behaviours remains true no matter the environment – including the tennis court. If a player is encouraged to perform a movement pattern in a certain way (i.e. backhand groundstroke) and it is not until their later years that they realise the technique of that movement pattern has been doing them more harm than good, they will be prompted to *learn* from this mistake. To correct the movement pattern itself, 'rewiring' is needed to re-learn and condition a new response – which can be quite lengthy, dependent on the 'entrenchment' of the technique, however, most players will need to go through this process at multiple *stages* before they even embark on the later end of the Pathway towards a Top 10 ranking.

> But imagine if you didn't need to go through this Level of 'rewiring' and you were equipped with the tools to progress towards a Top 10 ranking.

That's the power of the *Collection*. When using **My [Tennis] Black Book** you will notice *other* aspects of your game to either tighten, tweak and/or modify to conform to the rulers laid bare across the extensive body of work. Once you begin to integrate the 8 Keys into your game, you will become conscious of the Level of 'rewiring' that is specific to your Level of progress i.e. *your* capacity to leverage the kinetic chain and at what stage. Once the initial *key* foundations have been established the need to 'rewire' will no longer be of concern courtesy of your baseline being set. From here, you'll be equipped to progress at a more progressive rate per each new peak performance cycle, rather than regress due to the need to 'rewire' a habitual pattern, courtesy of putting in the *work* to establish the necessary foundations.

Of course, nothing worth doing is rarely easy and *why* only a small percentage of players progress each season – steadily, towards the Top 10. Of these players, 2% will cross the threshold to become barrier breakers each season and with such a slim margin, those who truly put in the work have the added advantage. With **My [Tennis] Black Book** at your disposal, you have the advantage to *take* – it is up to you, however, whether you're willing to learn from your mistakes, whether known and/or unknown, and to put in the work each week of *your* Black Book and progressively track towards the Top 10 – be it this season or soon *after* your initial decade of play.

Mistakes will happen to all players and yet the *key* differential between players is who's willing to *learn* from them in contrast to those who are content in overlooking the weaker aspects of their game. Whilst it does require a shift in mindset that is an underlying *key* of the best players in the world, there's a reason why only 8% of players will maintain their hold inside the Top 10 each season and yet that percentage will continues to narrow should the demand of *maintenance* not be upheld. The number of players equipped with the 8 Keys and comfortable to continuously learn – from micro percentages to micro inferences, these are the core drivers that separate even the best in the world to those with Grand Slam *wins*. It's the conscious habit of proactively learning irrespective of how small the mistake may be and/or seem, for the rewards are often tenfold.

To build a habitual process – a behaviour that is learned over time and is consistently conditioned until a state of autonomy is reached, this is where **My [Tennis] Black Book** comes into play as *your* Black Book provides a steadfast reminder to commit *and* complete. But that's one side of the behavioural process. The other is your own behaviour through play – are you calm under pressure, or do you immediately throw your racket when a point doesn't go your way? These are equal parts of the conditioning process and remain dependent on the reinforcement. A player will build the resilient attitude of a Top 10 player that behaves in a respectful manner – if exposed to positive reinforcements; or, a player may succumb to the pressures of play with consistent outbursts worthy of a toddler not getting their own way – temper tantrums, yelling to screaming or similar that can be surmised as disrespectful behaviour *if* this behaviour is not corrected and as such, reinforced.

Dependent on the behaviour of choice, be it *good* or *bad*, it is the conditioned behaviour of choice. To be fair, however, no player will behave like an upset toddler *unless* there's a form of positive reinforcement – void of a *behavioural* correction which can also be established through acceptance opposed to proactive discouragement of the behaviour. Similar conditioning can be applied to the resilient attitude – calm under pressure with a respectful attitude. If a positive reinforcer does not coexist with the favoured attitude, then the behaviour is of no guarantee to be consistently upheld during times of heightened pressure. A player, alike all individuals, remains susceptible to their environment and those in it. It is thereby *your* choice to learn the origin of these behaviours – why they have been and/or were encouraged and/or discouraged and for you to now knowingly choose the person and/or the type of player you *want* to be – you're in control.

It is important nonetheless to provide a pillar to base your decisions – be it **Federer**, **Nadal**, **Serena** and/or **Djokovic**. Irrespective of your age, every player is at the mercy of their development, from *learning* to ask for help to *learning* to understand hormonal imbalances and/or changes. Whilst every individual will go through their respective developmental *stage*, the core differences between players and their attitudes is reinforced through their environmental context that is not necessarily in their control. What is in your control is *your* choice – to learn from those who you aspire to *behave* like. This stems back to your own ability to perceive good from bad and those responsible for shaping your world view. Every person is susceptible to their surroundings and the player is no different. That said, if you are of age to know better, there are no excuses other than a conscious choice to do better.

> Most people do not go around yelling and screaming if something doesn't go their way and it is no different for the player – on and/or off the court. Poor behaviour is never okay and setting the bar high means you're taking responsibility for your own actions and are prepared to *learn* respectful behaviours to ensure these are ultimately reflected in the person you choose to become.

At the end of the day, everyone makes mistakes and it is how you react to those mistakes that makes the difference. If you're able to learn from your mistake on court – *why* it happened and its cause, your performance will evolve as you embark on that next peak performance. In this sense, mistakes are a constant that can be flipped upside down to reflect an ongoing learning curve that has the capacity to expedite your rate of progress and build towards that next new baseline.

Week 2 of Block 4 Tasks

> **Episode** to listen to: #15
> **Book Chapters:**
> **Level 1:** *The Secrets to Optimal Performance Success (2016)* – 2.8
> **Level 2**: *What is Your Game Missing? (2019a)* – Ranking 16 *and* 17
> **Level 3**: *The Science of Elite Performance (2017)* – Chapter 3: Communicating Skill Adaptation and Method of Feedback

Week 3 of Block 4 Tasks

> **Episodes** to listen to: #33 *and* #34
> **Book Chapters:**
> **Level 1:** *The Secrets to Optimal Performance Success (2016)* – 5.2 *and* 5.3
> **Level 2**: *What is Your Game Missing? (2019a)* – Ranking 18 *and* 18
> **Level 3**: *The Science of Elite Performance (2017)* – Chapter 3: Ball Dynamics and Point of Contact

Topic: Hydration as Fuel

When it comes to generalised advice on fluid replacement to 'fuel' your body in respect to *being* hydrated or replacing electrolytes, there are a myriad of options to choose from that saturate the *fuelling* market that come with their share of misconceptions. For this purpose, fuel refers to hydration and how various hydration methods also serve the purpose of fuelling the player for performance and/or recovery. It is of course important to highlight that these are generalised principles.

As a player develops and their game matures towards the elite, they will have vastly different fuelling needs in contrast to a player in the earlier years of their initial decade of play. Once the fundamentals around hydration are understood, however, a player will be better educated and in turn more informed about the fuel that is right for them and they can learn *how* their body responds to these fuel sources. At the elite Level of play, a player's fuelling can be incredibly specific to aide *individual* recovery and performance. This also comes down to precise measurements to specific ratio's that consider, but are not limited to, a player's energy expenditure, their body weight, the Level of perspiration to their collective fluid loss and energy deficit to actively replace these levels during play and then to further replace these levels during recovery. A player's fuel during play and recovery will be different due to the two different purposes.

> Education does become key to be able to differentiate between the two and your body's needs. But this will come with time and intuitively knowing *what* your body demands.

If you're beyond the developmental range, these are key learnings to take on board to better understand *how* your body responds to different fuel sources. If you're on the cusp of the elite or *later*, your time is *now* – fuelling for purpose is something you should be incredibly conscious of and if not, *now* is as good a time than ever. Your fuelling requirements will differ from one Level to the next due to the demands on your body – training load, rate of exertion to varied climates. For the player beyond the developmental range but not yet elite, most of these individual factors won't fluctuate too much until you begin to ascend closer towards the Top 100 when the cumulative demands will largely increase. But that doesn't mean *not* to pay close attention to your fuelling now – it is quite the opposite. For the player inside the Top 100 who is *tracking* towards the Top 50, the increased demands on your body are *just* beginning and the closer you edge towards the Top 10 these demands will continue to increase, and your rate of recovery will let you know if you're hydrating to fuel, or otherwise. But *learning* now can set you up for greater success, later.

When hydration is controlled accordingly, there will be limited lag – a limited drop of performance and you should be able to continue to train and/or play at your current peak baseline before reaching that *next* peak performance. Of course, this is dependent on what training cycle you're in and that you're adhering to the rulers of periodisation. To *optimally* recover, fuelling for purpose is a must but this lag is largely dependent on *knowing* what your body needs and taking action – independently and/or with your team. Without learning how your body responds, you're placing yourself at a disadvantage. By understanding *how* your body performs and *how* to maintain your hydration levels, this is the type of knowledge that delivers the edge.

A universal rule, albeit generalised, the rate of which a player perspires during play is of equal if not more of what needs to be replaced. But this is not to be mistaken for lack of perspiration equals lack of fluid replacement. Whether a clear sign (perspiration) is evident or not, other factors should be considered, including but not limited to, the duration of exertion to environmental conditions. From the Level of humidity to cooler climates or a higher altitude, perspiration may remain and/or change – players are not equally influenced at the same rate and as such, hydration is individual, but it does remain as a constant and universal *must* for all players. This is why as a generalised rule, a player's rate of perspiration is merely one guide of many and should not be used as an explicit rule as one player may perspire the same amount as another *but* over different time periods i.e. 20 minutes in contrast to 60 minutes – accounting for the individual is *key*.

Once you reach your initial years on Tour, each player will vary in their rate of perspiration, like their rate of progress, and the fuel they choose as their core hydrating source. This is largely preferential to each player with similarities between electrolytes and the right balance for their own fuelling needs. Each player will have their own set preferences, however, what matters most is the right hydration as fuel mix – in *liquid* form, with adequate intake during specified time windows to benefit either the players performance and/or recovery needs.

For most developmental players water is an adequate hydration source. That is until you reach the point of 'no return' and begin to outperform your natural resources and *need* fuel to perform. What matters here is both *knowing* and timing. Learning to understand your body and to *pre*-fuel prior to play is just as important as fuelling during play to prevent early onset of dehydration and subsequently underperform. This is something the best players in the world still need to consciously manage each time they step onto court and these 'habits' can be formed *now* if they do not already exist and *if* they do, use this as a review of your current hydration methods and if they can be further optimised. Ask yourself *what* fuel works best for your body and *why*, the frequency and/or quantities in need, and press further into exploring your *own* hydration needs and how they help or hinder your performance and/or recovery.

By staying ahead of your hydration curve, you can better prepare to avoid dehydration or falling short of that next intended peak due to limited fuelling resources. At the elite Level, your hydration demands will need to be replenished more frequently due to the greater load on the body. This means what an elite player consumes will vary from a player who is *not* yet working at their ultimate max capacity. And I can guarantee you that what fuels the best in the world is not soft drink, soda water, energy drinks and/or coconut water. What I can tell you is that each player will have a flavour preference that is simply a personal choice, but beyond that the *fuelling* mix (i.e. that includes a variety of electrolytes) is dependent on their own bodily demands.

> The biggest mistake *pre*-elites make is consuming something else – overusing energy drinks or similar when their resources are aplenty in contrast to the need to replenish and recover.

If there is one universal fuel source, however, it is water – the true hydration hero that cannot be replaced. Added to – yes, but replaced – no. Building the foundations of proper hydration and *learning* what your body *needs* and *how* it fuels your performance is absolutely fundamental. This learning is a habitual process that can be developed and *learned* over time – uniquely for *you*. When it comes to additional hydrating sources, whilst coconut water, soft drink, soda water or energy drinks are *not* ideal fuelling sources, a scientifically established optimal recovery drink has been identified as flavoured skim milk – a more optimal alternative for players in *addition* to their personalised mix at the elite Level of play. At the end of the day, hydration is a process to be *learned* and you are in control of learning what works best for *you* to maintain your resources and to prevent a state of depletion – *not* conducive to your recovery or performance let alone reaching that *next* baseline.

Week 4 of Block 4 Tasks

> **Episode** to listen to: #16
> **Book Chapters:**
> **Level 1:** *The Secrets to Optimal Performance Success (2016)* – 2.9
> **Level 2**: *What is Your Game Missing? (2019a)* – Ranking 19 *and* 19
> **Level 3**: *The Science of Elite Performance (2017)* – Chapter 3: External Contributing Factors to the Technique

KEY TAKEAWAYS

Episode Notes:

Chapter Notes:

Topic Notes:

GOALS

What I would like to accomplish and/or improve this week:

What I am working on:

TALK IT OUT

What I want to discuss with my coach this week (and note if I did discuss this and the outcome):

What I want to learn more about this week (and note if I did in fact learn more about this):

REFLECTION

Favourite part of the week I am most pumped for:

NOTES

PROGRESS

Top 3 superpower moments:

Any goal updates?

Top 3 key learnings to bring forward:

Biggest motivator:

1

TRAINING SESSION DIARY

Monday:

Tuesday:

Wednesday:

Thursday:

Friday:

Saturday/Sunday:

TOURNAMENTS

Tournament I'm playing and/or following:

Favourite player to watch in-action:

IMPROVEMENTS:

What I want to improve?

Biggest distraction?

How was my nutrition?

MATCHES/RESULTS

Opponent:

Tournament:

Score:

My Strength:

My Weakness:

Op Strength:

Op Weakness:

Other:

Opponent:

Tournament:

Score:

My Strength:

My Weakness:

Op Strength:

Op Weakness:

Other:

Opponent:

Tournament:

Score:

My Strength:

My Weakness:

Op Strength:

Op Weakness:

Other:

NON-TENNIS

Highlights:

Hobby I am most enjoying:

NOTES

MY NUMBERS FOR THE WEEK

Hours on Court:

Hours on Fitness:

Work Ethic / 10:

KEY TAKEAWAYS

Episode Notes:

Chapter Notes:

Topic Notes:

GOALS

What I would like to accomplish and/or improve this week:

What I am working on:

REFLECTION

Favourite part of the week I am most pumped for:

NOTES

TALK IT OUT

What I want to discuss with my coach this week (and note if I did discuss this and the outcome):

What I want to learn more about this week (and note if I did in fact learn more about this):

PROGRESS

Top 3 superpower moments:

Any goal updates?

Top 3 key learnings to bring forward:

Biggest motivator:

2

TRAINING SESSION DIARY

Monday:

Tuesday:

Wednesday:

Thursday:

Friday:

Saturday/Sunday:

TOURNAMENTS

Tournament I'm playing and/or following:

Favourite player to watch in-action:

IMPROVEMENTS:

What I want to improve?

Biggest distraction?

How was my nutrition?

MATCHES/RESULTS

Opponent:

Tournament:

Score:

My Strength:

My Weakness:

Op Strength:

Op Weakness:

Other:

Opponent:

Tournament:

Score:

My Strength:

My Weakness:

Op Strength:

Op Weakness:

Other:

Opponent:

Tournament:

Score:

My Strength:

My Weakness:

Op Strength:

Op Weakness:

Other:

NON-TENNIS

Highlights:

Hobby I am most enjoying:

NOTES

MY NUMBERS FOR THE WEEK

Hours on Court:

Hours on Fitness:

Work Ethic / 10:

My [Tennis] Black Book: Federer, Nadal, Serena, Djokovic, __________

KEY TAKEAWAYS

Episode Notes:

Chapter Notes:

Topic Notes:

GOALS

What I would like to accomplish and/or improve this week:

What I am working on:

REFLECTION

Favourite part of the week I am most pumped for:

NOTES

TALK IT OUT

What I want to discuss with my coach this week (and note if I did discuss this and the outcome):

What I want to learn more about this week (and note if I did in fact learn more about this):

PROGRESS

Top 3 superpower moments:

Any goal updates?

Top 3 key learnings to bring forward:

Biggest motivator:

3

TRAINING SESSION DIARY

Monday:

Tuesday:

Wednesday:

Thursday:

Friday:

Saturday/Sunday:

TOURNAMENTS

Tournament I'm playing and/or following:

Favourite player to watch in-action:

IMPROVEMENTS:

What I want to improve?

Biggest distraction?

How was my nutrition?

MATCHES/RESULTS

Opponent: ____________

Tournament: ____________

Score: ____________

My Strength: ____________

My Weakness: ____________

Op Strength: ____________

Op Weakness: ____________

Other: ____________

Opponent: ____________

Tournament: ____________

Score: ____________

My Strength: ____________

My Weakness: ____________

Op Strength: ____________

Op Weakness: ____________

Other: ____________

Opponent: ____________

Tournament: ____________

Score: ____________

My Strength: ____________

My Weakness: ____________

Op Strength: ____________

Op Weakness: ____________

Other: ____________

NON-TENNIS

Highlights:

Hobby I am most enjoying:

NOTES

MY NUMBERS FOR THE WEEK

Hours on Court:

Hours on Fitness:

Work Ethic / 10:

KEY TAKEAWAYS

Episode Notes:

Chapter Notes:

Topic Notes:

GOALS

What I would like to accomplish and/or improve this week:

What I am working on:

REFLECTION

Favourite part of the week I am most pumped for:

NOTES

TALK IT OUT

What I want to discuss with my coach this week (and note if I did discuss this and the outcome):

What I want to learn more about this week (and note if I did in fact learn more about this):

PROGRESS

Top 3 superpower moments:

Any goal updates?

Top 3 key learnings to bring forward:

Biggest motivator:

4

TRAINING SESSION DIARY

Monday:

Tuesday:

Wednesday:

Thursday:

Friday:

Saturday/Sunday:

IMPROVEMENTS:

What I want to improve?

Biggest distraction?

How was my nutrition?

MATCHES/RESULTS

Opponent:

Tournament:

Score:

My Strength:

My Weakness:

Op Strength:

Op Weakness:

Other:

Opponent:

Tournament:

Score:

My Strength:

My Weakness:

Op Strength:

Op Weakness:

Other:

Opponent:

Tournament:

Score:

My Strength:

My Weakness:

Op Strength:

Op Weakness:

Other:

TOURNAMENTS

Tournament I'm playing and/or following:

Favourite player to watch in-action:

NON-TENNIS

Highlights:

Hobby I am most enjoying:

NOTES

MY NUMBERS FOR THE WEEK

Hours on Court:

Hours on Fitness:

Work Ethic / 10:

End of Month Block Review & Additional Notes

KEY TAKEAWAYS

TALK IT OUT

GOALS

REFLECTION

TRAINING SESSION DIARY

NON-TENNIS

PROGRESS

IMPROVEMENTS

NOTES

MATCHES/RESULTS

Opponent: ____________________

Tournament: ____________________

Score: ____________________

My Strength: ____________________

My Weakness: ____________________

Op Strength: ____________________

Op Weakness: ____________________

Other: ____________________

Opponent: ____________________

Tournament: ____________________

Score: ____________________

My Strength: ____________________

My Weakness: ____________________

Op Strength: ____________________

Op Weakness: ____________________

Other: ____________________

Opponent: ____________________

Tournament: ____________________

Score: ____________________

My Strength: ____________________

My Weakness: ____________________

Op Strength: ____________________

Op Weakness: ____________________

Other: ____________________

May: Block 5

GOALS:

Week 1 of Block 5 Tasks

Episodes to listen to: #28 *and* #36
Book Chapters:
Level 1: *The Secrets to Optimal Performance Success (2016)* – 4.7 *and* 5.5
Level 2: *What is Your Game Missing? (2019a)* – Ranking 21 *and* 21
Level 3: *The Science of Elite Performance (2017)* – Chapter 4: Sport Psychology and Performance Outcomes

Topic: Taking Time Off

At times a player may indicate symptoms akin to being 'sick' when they are otherwise healthy. Symptoms typical of a cold and/or flu may present themselves in varying forms during times when the player has a vigorous and/or extensive training load with limited periods of rest – if at all. These are *key* indicators of over-training that should be taken seriously – taking time off should be encouraged with the necessary recovery protocols. Whilst general 'sickness' happens to the best of us, these symptoms can come about in otherwise healthy individuals when their training hasn't factored in rest periods during increased training loads.

Periodisation in its simplest form represents more concise *planning* to reach that next peak performance. These peak performances are oftentimes around *key* events and/or tournaments as a player's training is structured towards the player reaching an optimised 'State of Play' during the pre-*planned* window. Done right, periodisation can have incredible benefits and goes hand-in-hand with Grand Slam *success*. Nonetheless, some of the best players in the world succumb to the arduous continuation of the WTA and/or ATP Tours season with limited rest periods available throughout the calendar year. This is further amplified for a Top 10 player when their ranking correlates *with* their performance as deep runs mean limited rest opportunities between events.

For players who do *not* correlate with their ranking (at the higher echelon of play) and are *not* making successive deep runs, they will have more time off to rest before their next event. Ironically, the more a player does lose, the more time they have to rest and *plan* to peak at their next event. For the player who is competing in back-to-back events, sooner or later their body will break down and their performance will suffer if these periods of rest are not integrated irrespective of the schedule. This is why some of the best players in the world will regress in the rankings periodically to ensure their body can recover as neither the WTA nor ATP Tours scheduling are conducive to player recovery. This is something *you* must account for and understand – the trade-offs that are necessary at the top of the game to keep your body healthy.

> Your ranking will come, and it will go before you establish yourself at the top of the game. But your health should always come first. In time, you will secure enough points to maintain your hold inside the 8% *with* adequate rest periods across the season.

To *navigate* peak performances and rankings progressions, players need to take control of their schedule to ensure they are not overcommitting and can tactfully *plan* each respective peak throughout the season. Whilst this does come at a price with *less* ranking points to collect, too many events will ultimately leave a player susceptible to the pitfalls of over-training and your season will be cut short with no more points to acquire until you're back healthy. Fortunately, *your* Black Book has factored *micro*-rest periods into each training Block.

No player is immune from over-training, irrespective of their ranking, and the pre-*planning* of your own schedule is important – each tournament across the season (at the beginning of *your* Black Book) that will allow you to better *plan* each peak training Block. If you are a player with points to defend this season, you'll know where you need to 'collect' and *plan* to peak to ensure your ranking maintains a positive trajectory in contrast to regressing. For those who are ranked inside the Top 100, this becomes even more pivotal to *plan* accordingly to build on that momentum and edge closer towards the Top 50 and ultimately the Top 20. This is possible inside a season if the right planning is in place. For most, allow yourself two to three seasons to find that rhythm but in that second season with the respective planned peaks, you will edge closer towards the Top 10 *if* you *plan* meticulously.

For the Top 10 player, it is a little different. At this Level, you already have a set number of points that have allowed you to become a barrier breaker. You've done the work to get there but now, you need to do the *work* to *stay*. If you do not defend the points that allowed you to track towards the Top 10, you're more susceptible to regress. This is why establishing a firm baseline that is built upon – from one peak performance to the next, is critical in contrast to rapid-fire results without the underlying *work* as this places you in a more vulnerable position to lose your newfound rankings peak inside the space of the same season. To counter this, you must *plan* accordingly and know when to peak and when to rest – when to take a rankings hit and when to collect. If you're heavily reliant on Grand Slam performances, be sure to peak. If you're not reliant on Grand Slam performances – *plan* to peak anyway as the points will place you in a better position throughout the season. Players who *plan* to peak each Grand Slam and deliver – these are the players who correlate with the 8% whether now or in the future. This is a significant marker to work towards and integrate into your seasonal cycle – *plan* to succeed at the Grand Slam Level and your ranking *will* correlate.

The hardest part in any players training load is managing to compete in 'enough' tournaments to have the chance to secure that breakthrough *win*. Of course, Top 10 players aside, whether you're ranked inside the Top 100 or qualified for a Grand Slam or are inside the Top 30, a *win* against one of the best players in the world *or* a player much higher in the rankings, has the potential to secure you more points than anticipated – and *run* with it. But if you continue to play one event to the next chasing that breakthrough *win* and do not account for the necessitated rest, you will succumb to over-training. These are the players at the greatest risk of *falling* if they do not pace themselves in contrast to rapid-fire results which often come with a price *if* not managed accordingly.

By following The Long Game a player is set to progress towards the Top 10 over a set timeline whilst avoiding the pitfalls of over-training. There is a significant reason *why* many junior Grand Slam champions do not replicate their results at the senior Level and for those that do, it is often years in the making due to performative setbacks that includes time off Tour due to the rehabilitation process of one or more sustained injuries. To sidestep this disruption, learning to recognise the signs of over-training and understanding *how* to put the right tools in place to prevent over-training from halting your career, is fundamental.

Recognising the early signs of over-training becomes *key* for all players to ensure they know their body and its current threshold. Overtime, this threshold will increase but until then, it needs to be listened to and followed. If a player is otherwise healthy but feeling less than 100% and/or producing less than optimal performances for no apparent reason, modification becomes *key*. This includes touch-work that develops a player's hand-eye coordination, for example, that can be applied in varying settings and/or drills whilst lightening the training load yet keeps a player in training.

But it could be too late. Oftentimes the initial damage of over-training has started before a player will pay closer attention to the signs of 'wear and tear'. This underscores the importance of periodisation and a player genuinely learning the difference between being tired from training, for example, to what a 'pulled' muscle feels like and/or if they are lacking energy when they are otherwise healthy. If a player does not learn the initial signs of over-training, they are more susceptible to producing rapid-fire results – *before* injury catches them unaware.

Whilst there is quite a lot more to the topic of over-training, taking time off intermittently throughout the season will better protect you from its onset.

> Becoming familiar with periodisation is an incredibly rewarding experience for you to take control of your performance trajectory. Not only can it form an active *plan* committed towards by both player and coach, but it can also allow the parent and/or guardian an overview of what's happening, what's involved and points of discussion for the triangular relationship.

Periodisation actively encourages the planning of rest periods to balance out the respective peak performances throughout the season which also means you can *plan* for *fun* times outside of tennis. Keeping this balance is *key* and why each training Block inside **My [Tennis] Black Book** should include both higher and lower intensity weeks, and complete periods of rest to afford time to reflect on the Block you've just completed, the Block that's about to begin, and your goals for this season and *more*.

Week 2 of Block 5 Tasks

> **Episode** to listen to: #35
> **Book Chapters:**
> **Level 1:** *The Secrets to Optimal Performance Success (2016)* – 5.4
> **Level 2**: *What is Your Game Missing? (2019a)* – Ranking 22 *and* 23
> **Level 3**: *The Science of Elite Performance (2017)* – Chapter 4: The Role of the Tennis Racket and its Contribution to Player Performance

Week 3 of Block 5 Tasks

Episodes to listen to: #37 *and* #38
Book Chapters:
Level 1: *The Secrets to Optimal Performance Success (2016)* – 5.6 *and* 5.7
Level 2: *What is Your Game Missing? (2019a)* – Ranking 24 *and* 25
Level 3: *The Science of Elite Performance (2017)* – Chapter 5: Technology and Tennis

Topic: Injuries Correlating with Technique

Truth be told, one or more of your techniques may directly correlate with the *development* of an injury over the course of your career *if* not modified and/or amended specifically for *you*. It is that simple. Most players are taught a specific technique to 'play' the game and yet these techniques are one and the same in the onset of injury. Understandably, players inside the developmental spectrum will not notice an immediate discomfort due to the somewhat pliability of their growing bodies and oftentimes as they become conditioned to the technique they forgo any questioning of discomfort.

Once this developmental phase has run its course and a player is of the belief that their technique is the 'right' technique, any associated ailments are often overlooked as being attributed to their technique but rather a part of play and/or of the game. This is by no means true. It has been scientifically established that by introducing technical parameters that adhere to the properties of The V by Dr B © a player is not only safeguarding themselves from injury, but these techniques are directly correlated *with* Grand Slam success and Top 10 rankings. A staggering find.

Once a player is conditioned to perform a certain 'stroke' the warning signs are often overlooked due to a relative state of autonomy being reached. But if you know better, you can look to modify your technique to keep the *feel* whilst amplifying your results. Knowledge is power and by knowing the difference between a technique that correlates with injury and a technique that correlates with longevity inside the 8% will allow you to *plan* accordingly and modify your respective technical foundations over time – no matter your place on the performance spectrum.

The principles of human movement are readily available for all and help *navigate* the core principles that are responsible for the development of a respective discrete and/or serial skill. Unfortunately, these are often misguided and are not up to date with the latest scientific literature established throughout this extensive body of work. This is why the *Collection* has been made – for *you.* The recommended weekly readings are a mere starting point and share the tip of the iceberg to encourage you to make your way through the *Collection* over the course of several seasons to ensure what you read is retained and finds its way into *your* knowledge bank. The *Collection* is built as a learning curve to ensure a staggered by design approach – one building block after the next. Whilst the principles of human movement are a universal concept, the application of these to tennis have not previously been done, until the research behind the *Collection* established new technical findings, that in fact correlate *with* Top 10 rankings. Rather, the previous literature considered performance and varying outcomes but never was a complete technique considered with direct correlations identified to Grand Slam success and *more.* A similar discussion can be shared about how previous works did not account for the onset of injury if one or more discrete skills remained out of alignment.

A significant finding of this extensive research also identified the lack of a universal coach education system. Put simply, coaches are not 'educated' in the same way and/or agree on *key* principles that deliver optimised techniques and in turn, outcomes. How could they when this has never previously existed? Whilst there are various providers, educators to federations and/or similar across the globe, not one of them has access to nor shares the importance of the 8 Keys – this remains exclusive to the *Collection.* As such, no matter if a coach has been around for 50 years or 5 years, they are *not* being taught the 8 Keys which means they are unable to *teach* you the 8 Keys. This is why it is integral for all players when choosing their coach to ask the question — "Do you know the 8 Keys and if not, are you prepared to learn them and integrate them with me, into my game?"

> Due to the lapse of knowledge and human movement principles often being overlooked, injuries in players continue to halt their performance progressions and sometimes, permanently.

But this doesn't have to be the case for *you.* As the game has continued to reach new heights – more power and more speed demanded of the best in the world, techniques that were once thought to be 'good' are now the cause of players succumbing to their faults or rather at the mercy of simple human movement principles. Whether that be lack of rotation to remaining unfamiliar with the benefits of the kinetic chain, integrating functional movement patterns and more, if you're not actively learning and being mindful of *how* your game is progressing and *how* your body is responding, you're at the peril of your coach's knowledge base – absent of the 8 Keys unless they are predisposed to the 8 Keys. But if you *ask* the questions – as prompted throughout **My [Tennis] Black Book**, you're taking back control of your own technical foundations and together with your coach, you can learn the 8 Keys, you can learn the *key* technical parameters, and you can learn how to protect your body against injury whilst heading towards that next peak performance.

It often remains unknown to those involved at the developmental Level that injuries are being developed in conjunction with the conditioning of a new skill simply due to the lack of upkeep of *new* knowledge. Whether a discrete skill or a serial skill, if a technique is developed and does not follow the core principles of The V by Dr B © the player will be susceptible to injury whether in 6 weeks, 6 months or 6 years. A common example of an early onset injury is 'tennis elbow' which can easily come about by a player contacting the oncoming ball *behind* their elbow. It's that simple. The pressure placed on this joint becomes immense over time as the elbow is not designed to withstand this force. Similarly, a player may develop a lower back injury during their service motion if they are not activating their lower body or are yet to develop the technique in their lower body to counter the load placed on their upper body.

Irrespective if practicing your groundstroke or serve, the cause for concern is the Level of repetition of a compromised movement that is contributing to the development of 'tennis elbow' and/or a lower back injury. Over the course of a 30-minute session, a player may contact 100 balls – most definitely *more*, which equates to 100 x 'impacts' that are compromising *your* elbow and/or lower back as a bare minimum. If not corrected, over the course of 1 hour *five* days per week, that's 1,000 times your body is being compromised. You only need to do the math to see how quickly this number increases over the course of a training Block to notice how deeply entrenched some of these injuries can become when they go undetected.

To mitigate injuries, it is important for players to have a relative understanding of the *optimal* technique to ensure they ask questions when it comes to learning the respective discrete and/or serial skills and *how* they work to optimise their performance. Irrespective of *your* Level of play, injuries can happen at *all* levels and knowing what an optimised technique consists of can better prepare and protect you now and in the years to come. Whether you're a seasoned player on the WTA or ATP Tours, knowing the fundamentals behind your technique may be the difference between navigating towards a Top 10 ranking or falling short. It is never too late, however, to lessen the impact of a poor technique and you can start putting in the work, now.

This serves as a timely reminder for all that injuries can and do happen and often, they are due to poorly developed techniques, whether intentional or not. To safeguard yourself and your playing career, knowing the 8 Keys puts you ahead of the pack whilst sticking to completing *your* Back Book will ensure the *key* principles are learned and developed along the way – be it this season or as you progress through the respective *stages* and Levels. One thing is for sure, however, and that is optimised performances do not require injury at any stage of the Pathway, and you can put in the steps this season to optimally prepare to successfully *navigate* towards the Top 10 – this season or in the ones to come, without being sidelined along the way.

Week 4 of Block 5 Tasks

> **Episode** to listen to: #46
> **Book Chapters:**
> **Level 1:** *The Secrets to Optimal Performance Success (2016) – revisit* Chapter 1
> **Level 2**: *What is Your Game Missing? (2019a)* – Ranking 26 *and* 27
> **Level 3**: *The Science of Elite Performance (2017)* – Chapter 5: Movement during the Technique

MATCHES/RESULTS

Opponent:	Opponent:	Opponent:
Tournament:	Tournament:	Tournament:
Score:	Score:	Score:
My Strength:	My Strength:	My Strength:
My Weakness:	My Weakness:	My Weakness:
Op Strength:	Op Strength:	Op Strength:
Op Weakness:	Op Weakness:	Op Weakness:
Other:	Other:	Other:

KEY TAKEAWAYS

Episode Notes:

Chapter Notes:

Topic Notes:

GOALS

What I would like to accomplish and/or improve this week:

What I am working on:

TALK IT OUT

What I want to discuss with my coach this week (and note if I did discuss this and the outcome):

What I want to learn more about this week (and note if I did in fact learn more about this):

REFLECTION

Favourite part of the week I am most pumped for:

NOTES

PROGRESS

Top 3 superpower moments:

Any goal updates?

Top 3 key learnings to bring forward:

Biggest motivator:

1

TRAINING SESSION DIARY

Monday:

Tuesday:

Wednesday:

Thursday:

Friday:

Saturday/Sunday:

MY NUMBERS FOR THE WEEK

Hours on Court:

Hours on Fitness:

Work Ethic / 10:

IMPROVEMENTS:

What I want to improve?

Biggest distraction?

How was my nutrition?

MATCHES/RESULTS

Opponent:

Tournament:

Score:

My Strength:

My Weakness:

Op Strength:

Op Weakness:

Other:

Opponent:

Tournament:

Score:

My Strength:

My Weakness:

Op Strength:

Op Weakness:

Other:

Opponent:

Tournament:

Score:

My Strength:

My Weakness:

Op Strength:

Op Weakness:

Other:

TOURNAMENTS

Tournament I'm playing and/or following:

Favourite player to watch in-action:

NON-TENNIS

Highlights:

Hobby I am most enjoying:

NOTES

KEY TAKEAWAYS

Episode Notes:

Chapter Notes:

Topic Notes:

GOALS

What I would like to accomplish and/or improve this week:

What I am working on:

TALK IT OUT

What I want to discuss with my coach this week (and note if I did discuss this and the outcome):

What I want to learn more about this week (and note if I did in fact learn more about this):

REFLECTION

Favourite part of the week I am most pumped for:

NOTES

PROGRESS

Top 3 superpower moments:

Any goal updates?

Top 3 key learnings to bring forward:

Biggest motivator:

2

TRAINING SESSION DIARY

Monday:

Tuesday:

Wednesday:

Thursday:

Friday:

Saturday/Sunday:

TOURNAMENTS

Tournament I'm playing and/or following:

Favourite player to watch in-action:

IMPROVEMENTS:

What I want to improve?

Biggest distraction?

How was my nutrition?

MATCHES/RESULTS

Opponent:

Tournament:

Score:

My Strength:

My Weakness:

Op Strength:

Op Weakness:

Other:

Opponent:

Tournament:

Score:

My Strength:

My Weakness:

Op Strength:

Op Weakness:

Other:

Opponent:

Tournament:

Score:

My Strength:

My Weakness:

Op Strength:

Op Weakness:

Other:

NON-TENNIS

Highlights:

Hobby I am most enjoying:

NOTES

MY NUMBERS FOR THE WEEK

Hours on Court:

Hours on Fitness:

Work Ethic / 10:

KEY TAKEAWAYS

Episode Notes:

Chapter Notes:

Topic Notes:

GOALS

What I would like to accomplish and/or improve this week:

What I am working on:

TALK IT OUT

What I want to discuss with my coach this week (and note if I did discuss this and the outcome):

What I want to learn more about this week (and note if I did in fact learn more about this):

REFLECTION

Favourite part of the week I am most pumped for:

NOTES

PROGRESS

Top 3 superpower moments:

Any goal updates?

Top 3 key learnings to bring forward:

Biggest motivator:

3

TRAINING SESSION DIARY

Monday:

Tuesday:

Wednesday:

Thursday:

Friday:

Saturday/Sunday:

TOURNAMENTS

Tournament I'm playing and/or following:

Favourite player to watch in-action:

IMPROVEMENTS:

What I want to improve?

Biggest distraction?

How was my nutrition?

MATCHES/RESULTS

Opponent:

Tournament:

Score:

My Strength:

My Weakness:

Op Strength:

Op Weakness:

Other:

Opponent:

Tournament:

Score:

My Strength:

My Weakness:

Op Strength:

Op Weakness:

Other:

Opponent:

Tournament:

Score:

My Strength:

My Weakness:

Op Strength:

Op Weakness:

Other:

NON-TENNIS

Highlights:

Hobby I am most enjoying:

NOTES

MY NUMBERS FOR THE WEEK

Hours on Court:

Hours on Fitness:

Work Ethic / 10:

KEY TAKEAWAYS

Episode Notes:

Chapter Notes:

Topic Notes:

GOALS

What I would like to accomplish and/or improve this week:

What I am working on:

TALK IT OUT

What I want to discuss with my coach this week (and note if I did discuss this and the outcome):

What I want to learn more about this week (and note if I did in fact learn more about this):

REFLECTION

Favourite part of the week I am most pumped for:

NOTES

PROGRESS

Top 3 superpower moments:

Any goal updates?

Top 3 key learnings to bring forward:

Biggest motivator:

4

TRAINING SESSION DIARY

Monday:

Tuesday:

Wednesday:

Thursday:

Friday:

Saturday/Sunday:

MY NUMBERS FOR THE WEEK

Hours on Court:

Hours on Fitness:

Work Ethic / 10:

IMPROVEMENTS:

What I want to improve?

Biggest distraction?

How was my nutrition?

MATCHES/RESULTS

Opponent:

Tournament:

Score:

My Strength:

My Weakness:

Op Strength:

Op Weakness:

Other:

Opponent:

Tournament:

Score:

My Strength:

My Weakness:

Op Strength:

Op Weakness:

Other:

Opponent:

Tournament:

Score:

My Strength:

My Weakness:

Op Strength:

Op Weakness:

Other:

TOURNAMENTS

Tournament I'm playing and/or following:

Favourite player to watch in-action:

NON-TENNIS

Highlights:

Hobby I am most enjoying:

NOTES

End of Month Block Review & Additional Notes

KEY TAKEAWAYS | **GOALS**

TALK IT OUT | **REFLECTION**

TRAINING SESSION DIARY | **PROGRESS**

NON-TENNIS | **IMPROVEMENTS**

NOTES

MATCHES/RESULTS

Opponent:	Opponent:	Opponent:
Tournament:	Tournament:	Tournament:
Score:	Score:	Score:
My Strength:	My Strength:	My Strength:
My Weakness:	My Weakness:	My Weakness:
Op Strength:	Op Strength:	Op Strength:
Op Weakness:	Op Weakness:	Op Weakness:
Other:	Other:	Other:

June: Block 6

GOALS:

Week 1 of Block 6 Tasks

> **Episodes** to listen to: #41 *and* #42
> **Book Chapters:**
> **Level 1:** *The Secrets to Optimal Performance Success (2016)* – 6.3 *and* 6.4
> **Level 2**: *What is Your Game Missing? (2019a)* – Ranking 27 *and* 28
> **Level 3**: *The Science of Elite Performance (2017)* – Chapter 5: Athlete Perception and its Effect on Performance

Topic: Now and Future Planning

Generally speaking, when you've decided you are going to go 'somewhere' you either innately know how to get there (i.e. to the mall you've visited every week for the past 5 years) or if it is somewhere you're not as familiar with, you'll most probably ask Google, Siri, Alexa or similar with the help of Maps. With step-by-step directions provided with additional options presented along the way (i.e. do you need to stop for fuel?), you'll arrive at this 'somewhere' without much prior thought needed. Before Google, Siri and/or Alexa, among others, you had to rely on paper-form i.e. printed maps that were only accessible in larger books and then *plan* your journey before leaving whilst stopping at a few *key* points along the way to *check* your location. Quite outrageous and time consuming in contrast to now. But this kind of *planning* has evolved over the past 20 years yet there's still one primary form of planning that you genuinely need to be proactive about, quite simply because it is incredibly personal, and only *you* will have the answers.

To *plan* for the future, Google, Siri, Alexa and/or similar can help, as too can the various large language models (LLM) that are types of artificial intelligence (A.I). These LLM provide a new means to brainstorm – think 'aloud' whilst providing a 'tool' for you to engage with and even do the *planning* for you. But these LLM can't give you the answer you're after if you're not sure what question/s to ask. And the same applies in life itself. The human condition, however, can prompt you and guide you towards these questions to ensure you can learn what questions are the ones you seek the answers to, to better prepare you for your own planning.

But that sounds like a lot of work. Whether Google, Siri, Alexa and/or an A.I model (i.e. Claude, Pi, ChatGPT etcetera), it really comes down to personal preference – whether you do use these tools or not. The same applies to a more personal approach – interacting with others and asking the questions to the answers you seek outright to better ascertain varied approaches that were taken and/or ones to perhaps avoid. By collecting the information you're after from varied sources (i.e. not just one person and/or one tool), you're better able to gauge similar and/or differing interpretations to your question. More succinctly, when considering the notion of *planning* how one map may not be right for you, by asking the right personal questions for the answers you seek – what you want, where your 'somewhere' is, or how to go about getting there, this process *is* a form of planning and in turn you are ready to build out the map that you seek.

By using the planning analogy with maps and the various tools at your disposal to human interactions, *planning* for now and/or the future becomes streamlined with a little ingenuity. The adjacent *Journal* has been designed with you in mind, to *plan* at length and in detail – as you choose, whilst the integrated planning of *your* Black Book is more refined and will keep you accountable. To find the answers you seek, however, this is where your thoughts can run rampant – an adventure of their own, and your *Journal* can track each whirlwind until you arrive at your destination.

Whether now or for the future, *planning* for both forms a guidepost for you to follow. **My [Tennis] Black Book** is for now *with* future based contemplations. The future will vary from player to player with set commonalities i.e. Grand Slam success, but the road before this milestone can take varying shapes and forms from NCAA player to rising prodigy or late career *peak* performer.

> It's what happens now and the steps you put in action that will allow these future-based *plans* to have the greatest chance of becoming your reality.

If your 'somewhere' happens to be a Top 10 ranking, then the work you put in on a daily to weekly basis *now* matters. That's the built-in power of *your* Black Book to keep you on track. The same applies for fault-lines – *planning* is not always perfect. Life happens. Whether you're currently ranked inside the Top 50 on the WTA or ATP Tours or ranked inside the Top 10 at your local Academy – things won't always go to *plan*. Fault-lines will happen in your timeline – missed opportunities within the week or month, but **My [Tennis] Black Book** serves as your constant to lessen the blow and *learn* from these fault-lines – the choice will always be yours to make.

What's important to note about *planning* is its constancy and its change. You're allowed to change your metrics – whether that means less tournaments or one more. You're allowed to change your weekly on-court hours to flex with a weekend away to time with friends. You're allowed to make the time for *you* without sacrificing your end goal of a Top 10 ranking to a Grand Slam *win.* Although balance remains paramount. Every weekend can't be used to catch up on what's been missed *and* a thorough weekly reflection *and* to compete in a scheduled tournament. Balance matters and cramming is not the answer. By intentionally making the time to unwind, to spend time with those you cherish or to simply have time away, this too is a fundamental part of planning to ensure you have the time for the things that matter most. This is why every week of *your* Black Book your non-tennis highlights are a simple nudge – a gentle reminder of what you did this week for *you* away from the tennis court. But planning these moments from the beginning is *key* otherwise you'll become more susceptible to skipping these moments. Sure, one week is not an entire season, but habits start early and if not corrected, before you know it the end of the season may be around the corner, and you've only made a handful of moments just for *you.* Use this as your reminder to make that correction if in need.

> Follow **My [Tennis] Black Book** and you're halfway there. Use the nudges. Use the cues. The other half is up to you – *planning* is fundamental to *your* success and the tools are all inside, pointing in the same unified direction – towards the Top 10.

Whilst *planning* will look different for each player, it is about making the time in your day and across your week for what's either important or necessary. If you're required to attend classes in some form – secondary to tertiary, be sure to block this time off. If you have a preference to train in the afternoons – block this time off. If you have working responsibilities across the week – block these times off, too. If you have tasks to complete i.e. homework and/or dedicated study – block this time off daily and/or weekly. If you have specific strength and conditioning sessions multiple times per week – block these times off. If you go to bed at a certain hour to ensure you wake up at a certain time, again – block these times off. Of course, these are rudimentary examples and there are more complex tasks you may need to consider, but getting the foundations of your *planner* under control will allow you to build it into a more robust piece of art. Focus on your available time and integrate what you want to and/or what you should be doing, and proactively *plan* and use the earlier questions as a springboard as needed to help refine your *planner*.

Before you know it, you'll have a fully-fledged weekly *plan* that's all yours and that you're able to stick to, commit towards, and keep yourself accountable to on a rolling basis. Including the work of *your* Black Book plus external commitments, together you will have a more detailed 'life' *plan* accounting for both on and off court time to allow you to better manage your schedule and *plan* accordingly. This is an opportune time to draw on the adjoining *Journal* to include not only the details of *your* Black Book, but also the details of your external pursuits and/or commitments to keep them all inside the one safe place, easily accessible when in need.

Once your planning has considered both sides of the planning process, your future planning will be that much easier. The process of planning will become habitual over time after you commit to seeing it through. You'll also become more mindful of where improvements can be made, achieving a sounder balance to reaching a new performance baseline without too much afterthought. Remain constant and begin your planning now to ensure **My [Tennis] Black Book** becomes firmly integrated into your planning framework to steadily and progressively optimise your game, reach that next baseline, and edge you closer towards that future Top 10 ranking.

Week 2 of Block 6 Tasks

> **Episode** to listen to: #47
> **Book Chapters:**
> **Level 1:** *The Secrets to Optimal Performance Success (2016) – revisit* Chapter 2
> **Level 2**: *What is Your Game Missing? (2019a)* – Ranking 29 *and* 30
> **Level 3**: *The Science of Elite Performance (2017)* – Chapter 5: The Communication of Movement

Week 3 of Block 6 Tasks

Episodes to listen to: #43 *and* #58
Book Chapters:
Level 1: *The Secrets to Optimal Performance Success (2016), The Secrets to Optimal Coaching Success (2018)* – 6.5 *and* 2.2
Level 2: *What is Your Game Missing? (2019a)* – Ranking 30 *and* 31
Level 3: *The Science of Elite Performance (2017)* – Chapter 6: Technique and its Application to the Forehand

Topic: Your Kind of Happiness

What makes you happy? The 'real' type of happiness that causes your cheeks to hurt courtesy of the smirk that has taken over your face – can you name that kind of happiness? There is no right or wrong, just simply more of *that* kind of happiness is what it is all about. Akin to music to your ears, you're incredibly content with staying in your 'happy place' with no rush to leave and to simply enjoy the moment and the sensations of your own kind of happiness.

It may take a minute or two to find this kind of happiness if you do not visit this 'place' enough – genuinely putting the time aside to ask yourself what makes you happy. Whether a 'who' or a 'where' there's always a bonus etched in 'what' that's under your full control and readily available/accessible as you're the one in charge. If it is a 'who' then you might have to wait until that 'who' shows up (if not always next to you). Same goes for 'where' in case it is not a place and/or destination within walking distance. The kind of happiness you can easily control is a bonus. Of course, whether it is a 'what' or 'who' or 'where' to something in between, the primary topic here is about *your* happiness and how to set it off like fireworks on New Years Eve! The best part is that you never need to wait until that time of year since this kind of happiness – *your* kind of happiness, is somewhere your mind can shift to be in the here and now.

People are easy. Pets are, too – maybe even easier! The right kind of person and/or pet can set your smile alight with a simple hug. Where can be pretty easy, too. From a sunset or sunrise with the sand between your toes – you choose the destination, to a theme park or movie theatre as time simply stops when you're there as you immerse yourself in the experience. But what about the *what*? If you're thinking what I am thinking, there's a good chance it involves a yellow ball that sometimes gets pretty fuzzy but there's nothing like a fresh can and hitting the courts.

I am incredibly hopeful that tennis is a part of *what* makes you happy and brings you happiness with sheer joy every time you pick up your racket. It just feels right. It feels like home. It feels like a moment in time that can stop when it is you and that yellow ball and that *feeling* – the sensation that ever so slightly rattles through you with each impact. Rather, that moment in time when your strings make contact with that yellow ball and it is a nice and clean impact, as smooth and crisp as they come, timed perfectly – a kind of happiness that resonates with 'music to your ears'.

Nevertheless, this is not about tennis. Well, it is and is not. It's about finding what makes you happy other than tennis – a kind of happiness that may be on a similar Level, or completing different, but it is yours to call '*my* happy place'. Is it corny? Sure. Is it powerful with purpose? Absolutely. Although tennis may be a constant, it is just as important to enjoy yourself away from the court. By discovering other areas of life that bring you happiness – there's no right or wrong, it is simply *your* kind of happiness, this serves as a balancing act to 'diversify' your happiness to ensure it is always within reach. This is incredibly important for when things do not go to *plan*, be it back-to-back performances that were *less* than desired to increasing 'pressure to perform' or managing an injury.

> Either way, taking time away from tennis to reflect is incredibly important and in doing so, it is even more powerful to enjoy yourself away from the court — to thrive in another *kind* of happiness.

Another *kind* of happiness can be a superpower and a drawcard in and of itself. Not everyone will have this kind of happiness simply because they have not taken the time to ask themselves the oftentimes hard question. But it is there – all of you will have this kind of happiness – it is genuinely hidden if your focus hasn't deviated or that balancing act remains non-existent. Coexistence is the objective – for tennis and your 'other' kind of happiness to complement one another. Tennis is not designed to measure your self-worth nor your success as an individual or human being. Tennis is a sport at its basic Level and one that can be taken to new heights with the sheer love of the game. But tennis is *not* designed to be all-encompassing without the coexistence of balancing both kinds of happiness.

The beauty of various kinds of happiness is that they are all around you. Without being cliché there is a different kind of happiness around each new corner. The notion is to tap into this and remain conscious of oftentimes the smaller things as collectively multiple *kinds* of happiness can amass into something quite significant. And this is how your perspective can shift when the time is taken to take a step back from your immersion in the game you love and consider additional kinds of moments that bring you joy and make your cheeks hurt with laughter. This is when the real magic occurs. Call it a lightbulb moment or whatever you prefer, but taking a step back has a multitude of benefits that also serves as 'down time' to not simply keep that balance but also mitigate injuries caused from over-training.

Oftentimes players will not know they need to take a few days off – if not more. Your love for the game will keep you going coupled with your relentless determination to edge closer to that Top 10 ranking or be it Grand Slam *success*. But without the balance and without the different *kind* of happiness, your self-worth will become susceptible to your results and longevity at the top of the game will faulter as your inner-self becomes attached to your performance and your ability to enjoy the smaller things – find happiness away from the game, will eventually catch up to your ability to perform under pressure and pressure will *win*.

This is a double-edged sword when your love for the game can also be your downfall. But it doesn't have to be that way, *ever*. Balance is a powerful tool across a multitude of elements of the game and dispersed throughout *your* Black Book and when it comes to your happiness, it is literally a ruler to have *fun*.

> A player's happiness should never hinge on a result – in tennis or otherwise and if perchance your current happiness does rely on your performance and/or results, this is the time to begin to embrace additional *kinds* of happiness – friends, family, your dog and/or cat or reptile to bird – *whatever* garners that smirk. Go for it!

If you're still trying to figure out what may make *contribute* to your happiness, it may be as simple as a hike to a day out on the water – be it fishing and/or swimming or taking time to participate in a game of paintball/laser-tag to playing the most hyped multi-player online game. There are countless options at your disposal – take the time *now* to jot down a few options to consider this Block and next to ensure you have a number of *kinds* to trial and go ahead and find that sweet spot! Be it a trip to the museum or a new library, a book or a classic, a game of chess to painting or anything in between – the options are endless.

Happiness is subjective to everyone and there is zero correlation between a player's 'other' kind of happiness and their tennis success. Rather, the opposite is true – the intersection of both forms of happiness, on and off the court, is where the best players in the world take the advantage.

Week 4 of Block 6 Tasks

Episode to listen to: #48
Book Chapters:
Level 1: *The Secrets to Optimal Performance Success (2016) – revisit* Chapter 3 *and* Chapter 4
Level 2: *What is Your Game Missing? (2019a)* – Ranking 31 *and* 32
Level 3: *The Science of Elite Performance (2017)* – Chapter 6: Coach Communication and Athlete Responsiveness

MATCHES/RESULTS

Opponent:	Opponent:	Opponent:
Tournament:	Tournament:	Tournament:
Score:	Score:	Score:
My Strength:	My Strength:	My Strength:
My Weakness:	My Weakness:	My Weakness:
Op Strength:	Op Strength:	Op Strength:
Op Weakness:	Op Weakness:	Op Weakness:
Other:	Other:	Other:

MATCHES/RESULTS

Opponent:	Opponent:	Opponent:
Tournament:	Tournament:	Tournament:
Score:	Score:	Score:
My Strength:	My Strength:	My Strength:
My Weakness:	My Weakness:	My Weakness:
Op Strength:	Op Strength:	Op Strength:
Op Weakness:	Op Weakness:	Op Weakness:
Other:	Other:	Other:

KEY TAKEAWAYS

Episode Notes:

Chapter Notes:

Topic Notes:

GOALS

What I would like to accomplish and/or improve this week:

What I am working on:

REFLECTION

Favourite part of the week I am most pumped for:

NOTES

TALK IT OUT

What I want to discuss with my coach this week (and note if I did discuss this and the outcome):

What I want to learn more about this week (and note if I did in fact learn more about this):

PROGRESS

Top 3 superpower moments:

Any goal updates?

Top 3 key learnings to bring forward:

Biggest motivator:

1

TRAINING SESSION DIARY

Monday:

Tuesday:

Wednesday:

Thursday:

Friday:

Saturday/Sunday:

IMPROVEMENTS:

What I want to improve?

Biggest distraction?

How was my nutrition?

MATCHES/RESULTS

Opponent: ______

Tournament: ______

Score: ______

My Strength: ______

My Weakness: ______

Op Strength: ______

Op Weakness: ______

Other: ______

Opponent: ______

Tournament: ______

Score: ______

My Strength: ______

My Weakness: ______

Op Strength: ______

Op Weakness: ______

Other: ______

Opponent: ______

Tournament: ______

Score: ______

My Strength: ______

My Weakness: ______

Op Strength: ______

Op Weakness: ______

Other: ______

TOURNAMENTS

Tournament I'm playing and/or following:

Favourite player to watch in-action:

NON-TENNIS

Highlights:

Hobby I am most enjoying:

NOTES

MY NUMBERS FOR THE WEEK

Hours on Court:

Hours on Fitness:

Work Ethic / 10:

KEY TAKEAWAYS

Episode Notes:

Chapter Notes:

Topic Notes:

GOALS

What I would like to accomplish and/or improve this week:

What I am working on:

TALK IT OUT

What I want to discuss with my coach this week (and note if I did discuss this and the outcome):

What I want to learn more about this week (and note if I did in fact learn more about this):

REFLECTION

Favourite part of the week I am most pumped for:

NOTES

PROGRESS

Top 3 superpower moments:

Any goal updates?

Top 3 key learnings to bring forward:

Biggest motivator:

2

TRAINING SESSION DIARY

Monday:

Tuesday:

Wednesday:

Thursday:

Friday:

Saturday/Sunday:

IMPROVEMENTS:

What I want to improve?

Biggest distraction?

How was my nutrition?

MATCHES/RESULTS

Opponent:

Tournament:

Score:

My Strength:

My Weakness:

Op Strength:

Op Weakness:

Other:

Opponent:

Tournament:

Score:

My Strength:

My Weakness:

Op Strength:

Op Weakness:

Other:

Opponent:

Tournament:

Score:

My Strength:

My Weakness:

Op Strength:

Op Weakness:

Other:

TOURNAMENTS

Tournament I'm playing and/or following:

Favourite player to watch in-action:

NON-TENNIS

Highlights:

Hobby I am most enjoying:

NOTES

MY NUMBERS FOR THE WEEK

Hours on Court:

Hours on Fitness:

Work Ethic / 10:

KEY TAKEAWAYS

Episode Notes:

Chapter Notes:

Topic Notes:

GOALS

What I would like to accomplish and/or improve this week:

What I am working on:

REFLECTION

Favourite part of the week I am most pumped for:

TALK IT OUT

What I want to discuss with my coach this week (and note if I did discuss this and the outcome):

What I want to learn more about this week (and note if I did in fact learn more about this):

NOTES

PROGRESS

Top 3 superpower moments:

Any goal updates?

Top 3 key learnings to bring forward:

Biggest motivator:

3

TRAINING SESSION DIARY

Monday:

Tuesday:

Wednesday:

Thursday:

Friday:

Saturday/Sunday:

MY NUMBERS FOR THE WEEK

Hours on Court:

Hours on Fitness:

Work Ethic / 10:

IMPROVEMENTS:

What I want to improve?

Biggest distraction?

How was my nutrition?

MATCHES/RESULTS

Opponent:

Tournament:

Score:

My Strength:

My Weakness:

Op Strength:

Op Weakness:

Other:

Opponent:

Tournament:

Score:

My Strength:

My Weakness:

Op Strength:

Op Weakness:

Other:

Opponent:

Tournament:

Score:

My Strength:

My Weakness:

Op Strength:

Op Weakness:

Other:

TOURNAMENTS

Tournament I'm playing and/or following:

Favourite player to watch in-action:

NON-TENNIS

Highlights:

Hobby I am most enjoying:

NOTES

KEY TAKEAWAYS

Episode Notes:

Chapter Notes:

Topic Notes:

GOALS

What I would like to accomplish and/or improve this week:

What I am working on:

REFLECTION

Favourite part of the week I am most pumped for:

TALK IT OUT

What I want to discuss with my coach this week (and note if I did discuss this and the outcome):

What I want to learn more about this week (and note if I did in fact learn more about this):

NOTES

PROGRESS

Top 3 superpower moments:

Any goal updates?

Top 3 key learnings to bring forward:

Biggest motivator:

4

TRAINING SESSION DIARY

Monday:

Tuesday:

Wednesday:

Thursday:

Friday:

Saturday/Sunday:

TOURNAMENTS

Tournament I'm playing and/or following:

Favourite player to watch in-action:

IMPROVEMENTS:

What I want to improve?

Biggest distraction?

How was my nutrition?

MATCHES/RESULTS

Opponent:

Tournament:

Score:

My Strength:

My Weakness:

Op Strength:

Op Weakness:

Other:

Opponent:

Tournament:

Score:

My Strength:

My Weakness:

Op Strength:

Op Weakness:

Other:

Opponent:

Tournament:

Score:

My Strength:

My Weakness:

Op Strength:

Op Weakness:

Other:

NON-TENNIS

Highlights:

Hobby I am most enjoying:

NOTES

MY NUMBERS FOR THE WEEK

Hours on Court:

Hours on Fitness:

Work Ethic / 10:

End of Month Block Review & Additional Notes

KEY TAKEAWAYS

TALK IT OUT

GOALS

REFLECTION

TRAINING SESSION DIARY

NON-TENNIS

PROGRESS

IMPROVEMENTS

NOTES

MATCHES/RESULTS

Opponent:	Opponent:	Opponent:
Tournament:	Tournament:	Tournament:
Score:	Score:	Score:
My Strength:	My Strength:	My Strength:
My Weakness:	My Weakness:	My Weakness:
Op Strength:	Op Strength:	Op Strength:
Op Weakness:	Op Weakness:	Op Weakness:
Other:	Other:	Other:

July: Block 7

GOALS:

Week 1 of Block 7 Tasks

> **Episodes** to listen to: #44 *and* #59
> **Book Chapters:**
> **Level 1:** *The Secrets to Optimal Performance Success (2016), The Secrets to Optimal Coaching Success (2018)* – 6.6 *and* 2.3
> **Level 2**: *What is Your Game Missing? (2019a)* – Ranking 34 *and* 35
> **Level 3**: *The Science of Elite Performance (2017)* – Chapter 7: The Science of Knee Bend

Topic: Fair Play and Ethics

Everyone has played someone who has broken the rules. Everyone knows that feeling when their ball is called out by their opponent when it was clearly inside the line. And everyone knows what it *feels* like losing to that kind of opponent. It is not nice. And if you happen to not be 'everyone' – *you* are one of the lucky ones who has not played against someone who breaks the rules, then I hope this remains to be *true* and you continue to lead by example. For many of you who have, you are not alone by a long stretch and it is so incredibly important to share with you that you are on the right side of *optimal* performance practices and the right side of reaching that *next* peak performance cycle. How do I know? Quite simply, those who choose not to abide by the rules do not have the moral compass needed (i.e. internal ethics) to follow The Long Game and their unsportsmanlike conduct will catch up with them well before they have the chance to make their mark on the WTA or ATP Tours.

Ascending towards the top of the game takes *work*. A lot of work. The 3D's that I have honed into any player over a 20-year timespan – and counting, are none other than: **Discipline**, **Dedication** *and* **Determination**. It is that simple. You need the discipline to show up even when it gets hard – or you do not want to, not today. Discipline will allow you to see it through no matter what. You need the dedication to continue to fine-tune your craft. Dedication will allow you to continuously integrate the 8 Keys and continue to solidify them into your game like clockwork. Dedication is an art. And you need the determination to stay the course and not give up when it seems like all doors have closed. Determination will allow you to come back from match point down. Determination will see you rise early to get that extra hour in or train that little longer before the sun goes down. Together, the 3D's are a force to be reckoned with and as *promised* in the Foreword, if you are eyeing the Golden Slam to *the* Grand Slam, my *secrets* – including the 3D's, are now yours as shared throughout the *Collection*.

> This combination of the 3D's is lethal in the right hands. So lethal in fact that I would argue that they are habitual behaviours to be conditioned now, if not already, that will lead you towards the Top 10 with heightened success.

The differential between the 3D's and Grand Slams to *more* remains held in the 8 Keys and their constancy throughout the entirety of a season. But to get to this point, fair play and ethics are absolute – you *win* courtesy of the *work* opposed to the shortcuts, and when you lose you take the lessons from the loss and come back better next time. These are behaviours tied to your personal ethics and the 3D's will stand firm against adversity and *champion* ethics of the highest accord.

Do not settle. Do not compromise. Do not think less when you are *more*. Do not be afraid to *champion* ethics that resonate with you alike the 3D's – fair play will *win* every time. It is in your DNA but is still a choice – your choice, whether this *champion* is the person you strive to be. Those who break the rules and do not abide by fair play have already compromised their ethics and as such, the 3D's do not coexist. Do more. Do better and hold yourself to the highest standards alike those who have come before – **Federer**, **Nadal**, **Serena** and **Djokovic**, as next in line will need to follow suit.

Ethics is on full display when a player chooses to tell a lie or the truth. Was the ball inside or outside the line? Let your ethics choose. But do not be fooled – compromised ethics can run rampant due to the 3D's not yet coexisting. Whilst there is a clear right and wrong that highlights the importance of fair play for all, if a player's ethics are compromised, they may need a nudge in the right direction to help steer them in a more ethical direction. Nonetheless, if your ethics remain true, there is nothing like a little adversity during play to use as added fire to really bring the *win home*. More earnestly, it is important to report those who ignore ethical play to ensure our game can remain fair as it continues to withstand the advances of time.

And time is a powerful ally. The code of conduct exists to ensure this kind of behaviour is not allowed and as time has advanced, so too have the rules around play and the way they are controlled. At the elite Level of play, electronic line calling is a near constant as are referees to professional umpires and *more*. At the elite Level, the 3D's are a superpower – but the stakes are higher, it depends on whose might is greater than the rest. At the pre-elite levels of play, these 'reinforcements' are not always a constant nor a guarantee. It is at these levels of play, however, when players are at the greatest risk of exposure to unfair play and poor ethics. As such, it is even more of a challenge to still be true to your decision to enact the 3D's and remain true to your ethics when it is not a universal practice. These are the behaviours that *will* see you through, not merely to the Top 100, but afford you that added advantage to crack the Top 50 and together with the 8 Keys, the Top 10 will become a reality.

Collectively, fair play and ethics amplify a player's game. Those who play by the rules have nothing to fear. As an added advantage, these ingrained behaviours not merely *help* your performance but also ensure you are eligible to continue to play the game you have grown to love. Ethics goes together with fair play and upholding the values and rules of the sport is an ongoing practice you will be tested on regularly, whether directly or indirectly once you ascend closer towards the Top 10. From anti-doping protocols, regular testing to media commitments, your ingrained ethical and fair behaviours will allow you to continue to uphold the highest values of the sport, even when the pressure mounts, and continue to understand *why* these rules exist in the first place – to protect you and the integrity of the game.

By embodying the values of fair play, ethics, and the 3D's you are playing your part in raising the bar for *your* generation and those to come. Win or lose, the game is so much more than one peak performance cycle to the next. Whilst a part of the process, the mental fortitude upheld courtesy of your internal moral compass – underpinned by fair play and your ethics, is necessary to achieve a Level of constancy and longevity inside the Top 10. The more you integrate the 8 Keys and allow them to become a part of your *playing* pedagogy, the more you will reap the rewards – complemented by a moral compass that will guide you closer towards a Grand Slam to potentially *a* Golden Slam. But remember, constancy is *key* as are the 8 pillars held within.

Week 2 of Block 7 Tasks

Episode to listen to: #49
Book Chapters:
Level 1: *The Secrets to Optimal Performance Success (2016) – revisit* Chapter 5
Level 2: *What is Your Game Missing? (2019a)* – Ranking 35 *and* 36
Level 3: *The Science of Elite Performance (2017)* – Chapter 7: Balance and Performance

Week 3 of Block 7 Tasks

> **Episodes** to listen to: #50 *and* #51
> **Book Chapters:**
> **Level 1:** *The Secrets to Optimal Performance Success (2016), The Secrets to Optimal Coaching Success (2018) – revisit* Chapter 6 *and* 1.0
> **Level 2**: *What is Your Game Missing? (2019a)* – Ranking 37 *and* 37
> **Level 3**: *The Science of Elite Performance (2017)* – Chapter 8: Point of Contact

Topic: Fundamental Skills

Discrete and serial skills are absolute essentials for a player to be able to play the game and their respective advancements will coincide with the advancements of the players skill ability and in turn, Level of play. What is important to understand between this *skill* 'interplay' is that one is reliant on the other. A discrete skill is a 'part' of a serial skill, and a serial skill is dependent on multiple discrete skills to form a whole skill. Irrespective if you are developing 'a' part of a serial skill – a discrete skill or working on a better rhythm and/or execution of a serial skill through play – how multiple discrete skills work together, the skill development process is fundamental to play the game and *how* a player's Level of skill forms. Unfortunately, not all players will function with 'ease' when playing due to many of these respective skills *not* becoming embedded in their game from an earlier age.

This form of skill-neglect places a player at a disadvantage and under greater pressure as their game evolves due to complexities involved in developing specific fundamental skills in a player's later years. But this can be a blessing in disguise for some players as if they were in fact conditioned to play a certain way and their technique does not include specific skills nor is it conducive to their performance, a Level of rewiring is necessary irrespective of their Level of play to correct and develop the *missing* link.

> Habitual behaviours become conditioned in a player's game from their early years and this includes the process of skill acquisition. A skill that is learned and repeated becomes a known behaviour, for example, *how* to hit the ball. But if this habitual behaviour has *missed* a skill in the process, this is when the rewiring needs to take place.

Even more so, if a conditioned skill is identified as being detrimental to a player's game, this needs to be corrected sooner rather than later to prevent possible long-term damage. Unfortunately, due to the nonexistence of a universal coach education framework until the *Collection*, players have been more susceptible to *missing key* discrete skills in their technique development which are often not identified until the onset of injury, and only then forced to correct their skill which also comes with temporary regressive results until that skill can reach that new Level of autonomy and replace the former conditioned behaviour.

Interestingly, there is somewhat of a correlation between outdoor activities and natural skill development and the ease in which a player may then learn a new skill and ultimately technique. But as times have changed, so too has this natural development. In earlier generations most developing players were encouraged to play outside and inadvertently learn new skills without much afterthought. These skills would help foster *key* fundamental skills required for tennis that would in turn act as the building blocks for more advanced skills. Fast-forward to the current generation and not all players have the natural ability to transfer these skills purely because they are not being developed at the same rate. From catching with your non-dominant hand, standing on one foot with one eye closed *or* similar catching activities, for example, these are no longer commonplace activities in contrast to previous generations, and this has had a roll-on effect to the ease in developing a whole skill that has shifted the skill developmental process for players all around the world.

The core strategy that underpins the overall development of a serial skill is to fill the gap – develop the *discrete* skill/s that can typically be related in explanation and/or feel to a natural activity. But as this transference is limited now more than ever before, added work is needed to develop these skills before they can become integrated into the serial skill itself. For example, sitting low through motion (i.e. increased flexion of the lower body) would typically be readily available in players if they had experienced surfing, baseball and/or softball, soccer or played elementary games likened to tunnel ball, zigzag and/or a variety of other outdoor activities. One universal discrete skill across these sports is a lowered centre of gravity (C.O.G) with variations of lower body flexion depending on the sport and/or activity. Whilst all these sports require a myriad of discrete skills to perform and/or play them effectively, learning the respective fundamental skills are available in cross-sport/activity participation.

The advantage players once had simply due to playing outside was a powerful and yet inadvertent learning process. Learning to catch a ball *was* a rudimentary skill. Balancing on one foot *was* learned playing hopscotch. And coordination stemmed from hand-games to thumb-wars. All these activities may seem incredibly simple but in fact they are act as building block for larger skills – discrete and serial, and their respective advancements. This means if you are not playing outside and using this time to inadvertently learn skills that could very well help your performance, remember to make the time to *play* no matter your age.

In tennis, a player's coordination is constantly challenged. If you have never tried playing tennis with your hand *alone* – now is the time. You may recall handball in your younger years (or even present) and the direct crossover of this serial skill. The Level of hand-eye coordination for the tennis player only continues to evolve – from slower paced balls to eventually returning serves upwards of 200km/hr. This type of speed and the reaction needed – your Level of anticipation, is incredibly significant. But it all starts with the right building blocks to ensure when you are at this Level of play, your skill Level is at the ready and fine-tuned.

It is important here, irrespective of your current Level of play, to be mindful of the roll of *both* discrete and serial skills and *how* you can integrate these learnings into your practice sessions before your game. If working on loading the outside leg, dynamic balance will become *key* and playing points out 'hopping'– alternating legs, is a prime example of a 'building block' to help with this loading phase. This is one of many approaches that inadvertently builds coordination and strength of movement in tandem, but modifications of play to support and develop a skill is incredibly beneficial for your overall skill development and game.

Progressions from one *stage* to the next – without fast-forwarding, or missing a step, is *how* a player will hone their respective skills and avoid the need for rewiring in their later years. Many players on the WTA and ATP Tours have developed skills that continue to hold them back from ascending towards a Top 10 ranking. After all, there is a reason not every player breaks into the Top 100 let alone the Top 50 and beyond. These skills vary from poor technical mechanics that prevent a sounder service motion with an increased power output, to a lack of autonomy between their upper and lower body which causes several issues. Perhaps most importantly is the Level of compromise placed on a player's body at any given time when these shortcuts have been taken and *why* injuries are commonplace at both senior and junior levels of play.

The good news is that this does *not* need to be the case. Historically, injuries were never as high as they are now, but this is due to the advancements of the game i.e. increased speed and power, and the lack of technical player adaptations to be conditioned. Put simply, most players are still conditioned to play the game like those of earlier generations. Sure, the output of speed and power is a given, but *how* the body now needs to move, and the corresponding technical capacity, is *key*.

> Freedom to perform is developed over time and embedded in the integration of the 8 Keys. Taking into consideration *new* developments to account for *new* changes is a technical must for the player who aspires to follow in the footsteps of **Federer**, **Nadal**, **Serena** and/or **Djokovic**.

Sure, a player can in fact reach an elite Level of play with poor technical mechanics and compromised conditioning, *but* their ability to stay at the top of the game is not unanimous and their risk of injury is near absolute. A player with compromised skills that lacks the 'freedom to perform' – a fluid and rhythmic player in motion that embodies the 8 Keys, will remain challenged in their movements and their skill execution until time is taken to rewire and condition anew. The sooner a player learns the core building blocks, however, opposed to reaching a new Level of autonomy in an inferior and possibly detrimental skill, the sooner they will be back on track to steadily progress towards the elite Level of play *with* heightened rates of success.

Week 4 of Block 7 Tasks

> **Episode** to listen to: #60
> **Book Chapters:**
> **Level 1:** *The Secrets to Optimal Coaching Success (2018)* – 2.4
> **Level 2**: *What is Your Game Missing? (2019a)* – Ranking 38 *and* 39
> **Level 3**: *The Science of Elite Performance (2017)* – Chapter 8: Kinetic Chain *and* Force *and* Movement

KEY TAKEAWAYS

Episode Notes:

Chapter Notes:

Topic Notes:

GOALS

What I would like to accomplish and/or improve this week:

What I am working on:

REFLECTION

Favourite part of the week I am most pumped for:

TALK IT OUT

What I want to discuss with my coach this week (and note if I did discuss this and the outcome):

What I want to learn more about this week (and note if I did in fact learn more about this):

NOTES

Top 3 superpower moments:

Any goal updates?

PROGRESS

TRAINING SESSION DIARY

Monday:

Tuesday:

Wednesday:

Thursday:

Friday:

Saturday/Sunday:

Top 3 key learnings to bring forward:

Biggest motivator:

1

TOURNAMENTS

Tournament I'm playing and/or following:

Favourite player to watch in-action:

IMPROVEMENTS:

What I want to improve?

Biggest distraction?

How was my nutrition?

MATCHES/RESULTS

Opponent:

Tournament:

Score:

My Strength:

My Weakness:

Op Strength:

Op Weakness:

Other:

Opponent:

Tournament:

Score:

My Strength:

My Weakness:

Op Strength:

Op Weakness:

Other:

Opponent:

Tournament:

Score:

My Strength:

My Weakness:

Op Strength:

Op Weakness:

Other:

NON-TENNIS

Highlights:

Hobby I am most enjoying:

NOTES

MY NUMBERS FOR THE WEEK

Hours on Court:

Hours on Fitness:

Work Ethic / 10:

KEY TAKEAWAYS

Episode Notes:

Chapter Notes:

Topic Notes:

GOALS

What I would like to accomplish and/or improve this week:

What I am working on:

REFLECTION

Favourite part of the week I am most pumped for:

TALK IT OUT

What I want to discuss with my coach this week (and note if I did discuss this and the outcome):

What I want to learn more about this week (and note if I did in fact learn more about this):

NOTES

PROGRESS

Top 3 superpower moments:

Any goal updates?

Top 3 key learnings to bring forward:

Biggest motivator:

2

TRAINING SESSION DIARY

Monday:

Tuesday:

Wednesday:

Thursday:

Friday:

Saturday/Sunday:

TOURNAMENTS

Tournament I'm playing and/or following:

Favourite player to watch in-action:

IMPROVEMENTS:

What I want to improve?

Biggest distraction?

How was my nutrition?

MATCHES/RESULTS

Opponent: ----------

Tournament: ----------

Score: ----------

My Strength: ----------

My Weakness: ----------

Op Strength: ----------

Op Weakness: ----------

Other: ----------

Opponent: ----------

Tournament: ----------

Score: ----------

My Strength: ----------

My Weakness: ----------

Op Strength: ----------

Op Weakness: ----------

Other: ----------

Opponent: ----------

Tournament: ----------

Score: ----------

My Strength: ----------

My Weakness: ----------

Op Strength: ----------

Op Weakness: ----------

Other: ----------

NON-TENNIS

Highlights:

Hobby I am most enjoying:

NOTES

MY NUMBERS FOR THE WEEK

Hours on Court:

Hours on Fitness:

Work Ethic / 10:

KEY TAKEAWAYS

Episode Notes:

Chapter Notes:

Topic Notes:

GOALS

What I would like to accomplish and/or improve this week:

What I am working on:

REFLECTION

Favourite part of the week I am most pumped for:

TALK IT OUT

What I want to discuss with my coach this week (and note if I did discuss this and the outcome):

What I want to learn more about this week (and note if I did in fact learn more about this):

NOTES

PROGRESS

Top 3 superpower moments:

Any goal updates?

Top 3 key learnings to bring forward:

Biggest motivator:

3

TRAINING SESSION DIARY

Monday:

Tuesday:

Wednesday:

Thursday:

Friday:

Saturday/Sunday:

TOURNAMENTS

Tournament I'm playing and/or following:

Favourite player to watch in-action:

IMPROVEMENTS:

What I want to improve?

Biggest distraction?

How was my nutrition?

MATCHES/RESULTS

Opponent:

Tournament:

Score:

My Strength:

My Weakness:

Op Strength:

Op Weakness:

Other:

Opponent:

Tournament:

Score:

My Strength:

My Weakness:

Op Strength:

Op Weakness:

Other:

Opponent:

Tournament:

Score:

My Strength:

My Weakness:

Op Strength:

Op Weakness:

Other:

NON-TENNIS

Highlights:

Hobby I am most enjoying:

MY NUMBERS FOR THE WEEK

Hours on Court:

Hours on Fitness:

Work Ethic / 10:

NOTES

My [Tennis] Black Book: Federer, Nadal, Serena, Djokovic, ________

KEY TAKEAWAYS

Episode Notes:

Chapter Notes:

Topic Notes:

GOALS

What I would like to accomplish and/or improve this week:

What I am working on:

TALK IT OUT

What I want to discuss with my coach this week (and note if I did discuss this and the outcome):

What I want to learn more about this week (and note if I did in fact learn more about this):

REFLECTION

Favourite part of the week I am most pumped for:

NOTES

PROGRESS

Top 3 superpower moments:

Any goal updates?

Top 3 key learnings to bring forward:

Biggest motivator:

4

TRAINING SESSION DIARY

Monday:

Tuesday:

Wednesday:

Thursday:

Friday:

Saturday/Sunday:

TOURNAMENTS

Tournament I'm playing and/or following:

Favourite player to watch in-action:

IMPROVEMENTS:

What I want to improve?

Biggest distraction?

How was my nutrition?

MATCHES/RESULTS

Opponent:

Tournament:

Score:

My Strength:

My Weakness:

Op Strength:

Op Weakness:

Other:

Opponent:

Tournament:

Score:

My Strength:

My Weakness:

Op Strength:

Op Weakness:

Other:

Opponent:

Tournament:

Score:

My Strength:

My Weakness:

Op Strength:

Op Weakness:

Other:

NON-TENNIS

Highlights:

Hobby I am most enjoying:

NOTES

MY NUMBERS FOR THE WEEK

Hours on Court:

Hours on Fitness:

Work Ethic / 10:

End of Month Block Review & Additional Notes

KEY TAKEAWAYS

TALK IT OUT

GOALS

REFLECTION

TRAINING SESSION DIARY

NON-TENNIS

PROGRESS

IMPROVEMENTS

NOTES

MATCHES/RESULTS

Opponent:	Opponent:	Opponent:
Tournament:	Tournament:	Tournament:
Score:	Score:	Score:
My Strength:	My Strength:	My Strength:
My Weakness:	My Weakness:	My Weakness:
Op Strength:	Op Strength:	Op Strength:
Op Weakness:	Op Weakness:	Op Weakness:
Other:	Other:	Other:

August: Block 8

GOALS:

Week 1 of Block 8 Tasks

Episodes to listen to: #52 *and* #61
Book Chapters:
Level 1: *The Secrets to Optimal Coaching Success (2018)* – 1.1 *and* 3.0
Level 2: *What is Your Game Missing? (2019a)* – Ranking 39 *and* 40
Level 3: *The Science of Elite Performance (2017)* – Chapter 8: Feedback Methods

Topic: Problem Solving

The ability to solve problems is not anything new but the ability to do so rapid-fire, in milliseconds under pressure and when in motion, is an artform. This artform does not happen overnight and is not a constant amongst the best players in the world. There is a *key* reason why some players have an innate ability to *know* when on court and under pressure – as they unravel the problem before them in real-time, *what* their next move needs to be and the one after that, until the point has been deconstructed and come to a close. And there is a reason this is not commonplace amongst all players inside the Top 100 let alone the Top 50 or Top 30. To overcome this immense hurdle and break the Top 20 benchmark there is a quintessential Level of problem solving that needs to reach a certain Level of autonomy *before* the Top 10 comes within reach.

But there is a lot of work to do, irrespective of what your current ranking is, to reach this state of autonomy and it begins with the habitual conditioning of solving multifaceted problems – *one* at a time. Not until your second decade of play will you begin to fine-tune these skills, but to get to this point of *knowing* requires both on and off court challenges as well as an innate understanding that has evolved over time, of the *game* you are playing.

Without the work and time spent knowing, understanding to conditioning the *how* and in varied scenarios, including but not limited to, under pressure, under time constraints and under set conditions, your ability to solve problems will fall short of that of **Federer**, **Nadal**, **Serena** or **Djokovic**. The intentional *work* needs to be done. What becomes most pertinent in any players development of their problem solving skills is their intuition which can be conditioned over time by making the time to study the game being played and putting their decisions into practice.

From knowing *when* the ball is coming to *where* the ball is going and *how* before you are a step ahead, to ensure you stay in control of the point in play, that's half the problem. The other problem is *knowing* what is happening on both sides of the court and preparing your response in those milliseconds. Over time, you will develop habitual behaviours to *optimally* respond to varied scenarios and the respective problems to be solved, creating your own internal action-reaction response and in turn, feedback loop. As you progress, this feedback loop will continue to expand and the more you come to understand the intricacies of the game, the more versatile your responses will become. But the real problem is ongoing. A new Level of problem solving at an autonomous Level is that next step as you edge closer towards the Top 10 and with that comes a multidimensional feedback loop that understands the varied outcomes of your own game *and* your opponents. When this milestone occurs, you'll know. And when it does, you'll know that Grand Slam awaits.

To reach this Level of autonomy the 7 Keys are a prerequisite and the more they become solidified in your game, the greater the power of your feedback loop. But remember – the 8th Key *will* keep you inside the Top 10 so long as all other Keys stay intact. And armed with your feedback loop *and* the 8 Keys is when the real *fun* begins.

> The game of tennis is so incredibly challenging and yet so rewarding. The complexities of the game can be often overlooked and this is where mistakes are readily made.

By not paying attention to the intricacies of the *how* your feedback loop will remain incomplete. As such, your ability to solve problems in milliseconds will become delayed – time that you do not have. Problem solving is an artform for a reason – it takes practice, a lot of work *and* developed intuition. At the elite Level of play, players who do not develop their anticipation, let alone perception will remain stuck outside the Top 50 at best. For inroads to be made, *how* to solve problems needs to reach that next state of autonomy for a player to reach that *next* peak performance when they are at these heights of the game.

The pressing question is *how*? Unfortunately, there is no universal answer yet there is a *key* difference and commonality between those who excel and become the best in the world to those who aspire to reach that Level but continue to fall short. The art is in the *knowing* and in the *understanding*. This is where the role of anticipation comes into play coupled with explosive power and speed. Power and speed are only half as good *if* anticipation lacks, and anticipation is only half as good *if* a player does not have the speed and power to overcome their opponents' Level of play. Neither happens without putting in the work and the time. This combination delivers a Top 10 status and is what the *final* stages of Grand Slams are made of.

Anticipation is knowing. Anticipation is understanding. Anticipation considers time and reaction to speed and execution – *then* it does it again on repeat. Whether the scoreline is 15-0 to 40-0 or 15-15 to advantage receiver will depend on how the point plays out. A player must *play* to figure out in real-time *how* one ball, at speed and in motion, will place them in a winning position. Whether two balls or twelve, a player needs to solve a live problem during play and their feedback loop is their arsenal to call upon to deconstruct the point and make the next winning move. This is where a player's Level of perception becomes *key*.

Together, anticipation and perception are all a part of the game and equally a part of solving real-time problems. How a player will react and respond is ultimately dependent on their Level of perception – what they perceive is happening on their side of the court, and also their opponents side of the court, then to call upon their action-reaction feedback loop to solve one ball at a time before the point comes to an end – by winner, forced or unforced error.

During a single point you are solving what is the *next* best move to then anticipate what your opponent will do *next* with both your perception and anticipation on full alert in the milliseconds before the next ball arrives and the next decision is made – your next problem to solve. And this is all happening within a single point. To reach this Level of autonomy at the highest Level, precise conditioning becomes *key* which can be complemented by inadvertent skill development and studying *how* your opponent plays the game to *plan* in advance. This is where *your* Black Book can further be leveraged to record your respective results to more accurately *plan* in the future. Furthermore, as you put the work in under different constraint and put your decision making into practice, your ability to solve problems more rapidly will evolve to reach that next 'State of Play'.

For a player to round out their arsenal of problem-solving skills, their reaction to response ratio needs to be finetuned. In conjunction with your feedback loop, the reaction-response feeds directly into your action-response. Whilst one emphasises the ability to act autonomously in a conditioned way, the other talks to your *re*-activity that may not be your intended action. But by conditioning your responses these two will become one and the same – intentional responses you're consciously in control of during play.

Tennis is incredibly multifaceted and problem solving is an immensely detailed and broad topic to accommodate and understand. By putting in the *work* and building the initial foundations of *how* – knowing, understanding and proactively studying the game, and your opponents, this will allow your natural intuitive process to take shape. In time, from one game to one season to the next, you will progressively develop your anticipatory skills and learn how to more effectively perceive play and ultimately solve problems. So when the time comes and you notice that your response ratio has become one and the same, you'll look back and understand the artform that underpins the game and *how* the feedback loop really is continuous as you look to catch that next new peak wave.

Week 2 of Block 8 Tasks

Episode to listen to: #62
Book Chapters:
Level 1: *The Secrets to Optimal Coaching Success (2018)* – 3.1
Level 2: *What is Your Game Missing? (2019a)* – Ranking 41 *and* 41
Level 3: *The Science of Elite Performance (2017)* – Chapter 8: The Coach Learning / Education Process

Week 3 of Block 8 Tasks

Episode to listen to: #53
Book Chapters:
Level 1: *The Secrets to Optimal Coaching Success (2018)* – 1.2
Level 2: *What is Your Game Missing? (2019a)* – Ranking 42 *and* 44
Level 3: *The Science of Elite Performance (2017)* – Chapter 8: Education and Sport

Topic: Reading and Performance

Reading is one of the oldest forms of learning – visual or otherwise, where you are engaging in a form of 'discovery' to learn something new. Whilst not all reading material is perceived as a form of learning, it remains contextual and dependent on the individual. Whether that be a comic book or textbook, the content will differ, but the underlying message of discovery remains. Interestingly, how you integrate reading into your life has various implications, including your *own* performance.

Over the years, reading has adapted to accommodate varied learning needs and as such is no longer exclusively in written form. The same applies for visual and how visual forms of reading (i.e. infographics) have become more commonplace. Auditory reading is a new form of reading – whether listening to a book and/or podcast **Episode**. The commonality is the element of discovery whereby there is a chance to *learn* something new in not one, but multiple forms – written, visual and/or auditory. These forms and/or shapes of reading can be deduced to a form of learning, whether intended or not.

When reading in written form the eye catches the detail of the words – the use of language and *how* it conveys its message. Fictional or non-fictional, the language stays as does the use of words and their meaning. Whatever the style you read is more likely to inadvertently reflect and influence your own style of writing over time. Similarly, visual forms of reading will tend to reflect and/or influence your own visual interpretations but without written words, you will miss the influence of language. Nonetheless, individuals learn in varying ways and whilst written form is more widespread, visual representations can be incredibly powerful, especially in the form of an infographic or *plan* that can catch your attention. When it comes to auditory, this is close to written form but with the caveat of a lack of visual – actively reading word for word. What needs to be considered here is the personal processing of words – of language, as you read word for word in contrast to taking a fast-tracked approach.

> The psychology behind human behaviour and how you learn varies from one person to the next and this is an element, along with *discovery* that is embedded in the 8 Keys.

It has been well documented that the ability of an individual to retain information is typically done so in written form and aides in their ability to learn. Whilst a case could be made for visual and/or auditory, due to written form being the primary learning source, it simply outnumbers other forms. Writing your own notes with pen and/or pencil and paper, for example, has been shown to increase an individual's Level of retention i.e. memory. Whether in a classroom, lecture hall or otherwise, the simple task of manually taking notes has a *feel* about it, different to that of typing, and whilst notes in typed form are beneficial, it is the *feeling* behind writing your own words as the graphite and/or ink touches the paper and your attention is present on writing each letter of each word as you etch your thoughts down onto the page. As such, using the written form of reading as a primary learning tool is still the best documented practice, but combining written form with both visual and auditory delivers a multi method approach to enhance your learning further and why this design has been integrated into *your* Black Book.

This mixed approach is accessible every week over the course of your season right *here*. If you are reading this with your own eyes and/or following along through an auditory form, the written form of reading is underway. Then comes the visual form – every week you have *plans* to complete. As for the *specific* auditory form, that is what the weekly Podcast **Episodes** are for – another learning form made available for you. Whilst not all information you read – whether written, visual or auditory, will be retained, by integrating this multi method approach – interchanging these modes of discovery learning with your own *notes* to complement your learning, affords a win-win to advance your performance.

It can't be stressed enough, however, that these forms of learning and the process of discovery is by human nature a skill in itself and a powerful one at that. The skill of reading and in turn an openness to learn is a must if the 8 Keys are to become integrated into your game. The more you read in written form, the easier it will become. The more you learn through visual forms, the easier it will become. And the more you read auditory form (i.e. following the narrator sentence by sentence), the easier it will also become, although should you choose not to follow the narrator, you are simply missing the *dual*-learning possibility.

There is no right or wrong but simply what works best for you. This is not to say if you do not like one format not to try it or to persevere, it means that you will innately know how you learn and by what form and/or shape of reading comes more naturally to you. It is important, nonetheless, not to distance yourself from classic form – reading word for word, simply because it may take longer than visual form, or to not occasionally try auditory form to reinforce what you have already learned. There is a complementary nature between all forms and why **My [Tennis] Black Book** puts this into practice for you to get the most out of your reading across varied forms.

When it comes to your performance, reading can be a means to *discovery*. By using this approach, reading can influence the perception of your surroundings

– your Level of awareness on court and what is the next move you will make. Whilst this may seem like a stretch of the imagination, when reading is viewed as a part of the discovery process, ask yourself – 'what am I learning that I can consider applying, today?' After all, learning and discovery go hand in hand. When you learn something new there is an element of discovery behind what you have just learned. For each new learning can be considered a discovery itself. In tennis, this discovery process is transferable if you reframe your view of reading and *how* learning through reading can in fact help your performance. Of course, there is an art to this discovery process but by staying alert – on the lookout for what you will learn *next* between the pages, this can have a compounding effect when you step onto the court.

In action, there is always something to be learned in some way, shape or form from your performance. By conditioning your mind through reading that the art of discovery is present through *play*, this is where you get to have fun using your reading to advance your performance more rapidly. Not only can the simple act of reading help you search for on-court clues – *what* ball is coming next or *what* is your opponent's most likely next move, looking for the learning curve during play can help you stumble across new approaches, new tools and new ways of playing, all from switching your learning mode to discovery. And it starts with reading, the willingness to learn and the openness to both write, listen and see more than first meets the eye.

Week 4 of Block 8 Tasks

> **Episode** to listen to: #5
> **Book Chapters:**
> **Level 1:** *The Secrets to Optimal Performance Success (2016)* – 1.5
> **Level 2**: *What is Your Game Missing? (2019a)* – Ranking 45 *and* 46
> **Level 3**: *The Science of Elite Performance (2017)* – Chapter 9: Coaching History

MATCHES/RESULTS

Opponent:
Tournament:
Score:
My Strength:
My Weakness:
Op Strength:
Op Weakness:
Other:

Opponent:
Tournament:
Score:
My Strength:
My Weakness:
Op Strength:
Op Weakness:
Other:

Opponent:
Tournament:
Score:
My Strength:
My Weakness:
Op Strength:
Op Weakness:
Other:

MATCHES/RESULTS

Opponent:
Tournament:
Score:
My Strength:
My Weakness:
Op Strength:
Op Weakness:
Other:

Opponent:
Tournament:
Score:
My Strength:
My Weakness:
Op Strength:
Op Weakness:
Other:

Opponent:
Tournament:
Score:
My Strength:
My Weakness:
Op Strength:
Op Weakness:
Other:

MATCHES/RESULTS

Opponent:
Tournament:
Score:
My Strength:
My Weakness:
Op Strength:
Op Weakness:
Other:

Opponent:
Tournament:
Score:
My Strength:
My Weakness:
Op Strength:
Op Weakness:
Other:

Opponent:
Tournament:
Score:
My Strength:
My Weakness:
Op Strength:
Op Weakness:
Other:

KEY TAKEAWAYS

Episode Notes:

Chapter Notes:

Topic Notes:

GOALS

What I would like to accomplish and/or improve this week:

What I am working on:

REFLECTION

Favourite part of the week I am most pumped for:

TALK IT OUT

What I want to discuss with my coach this week (and note if I did discuss this and the outcome):

What I want to learn more about this week (and note if I did in fact learn more about this):

NOTES

PROGRESS

Top 3 superpower moments:

Any goal updates?

Top 3 key learnings to bring forward:

Biggest motivator:

1

TRAINING SESSION DIARY

Monday:

Tuesday:

Wednesday:

Thursday:

Friday:

Saturday/Sunday:

TOURNAMENTS

Tournament I'm playing and/or following:

Favourite player to watch in-action:

IMPROVEMENTS:

What I want to improve?

Biggest distraction?

How was my nutrition?

MATCHES/RESULTS

Opponent:

Tournament:

Score:

My Strength:

My Weakness:

Op Strength:

Op Weakness:

Other:

Opponent:

Tournament:

Score:

My Strength:

My Weakness:

Op Strength:

Op Weakness:

Other:

Opponent:

Tournament:

Score:

My Strength:

My Weakness:

Op Strength:

Op Weakness:

Other:

NON-TENNIS

Highlights:

Hobby I am most enjoying:

NOTES

MY NUMBERS FOR THE WEEK

Hours on Court:

Hours on Fitness:

Work Ethic / 10:

KEY TAKEAWAYS

Episode Notes:

Chapter Notes:

Topic Notes:

GOALS

What I would like to accomplish and/or improve this week:

What I am working on:

REFLECTION

Favourite part of the week I am most pumped for:

TALK IT OUT

What I want to discuss with my coach this week (and note if I did discuss this and the outcome):

What I want to learn more about this week (and note if I did in fact learn more about this):

NOTES

PROGRESS

Top 3 superpower moments:

Any goal updates?

Top 3 key learnings to bring forward:

Biggest motivator:

2

TRAINING SESSION DIARY

Monday:

Tuesday:

Wednesday:

Thursday:

Friday:

Saturday/Sunday:

TOURNAMENTS

Tournament I'm playing and/or following:

Favourite player to watch in-action:

IMPROVEMENTS:

What I want to improve?

Biggest distraction?

How was my nutrition?

MATCHES/RESULTS

Opponent:
Tournament:
Score:
My Strength:
My Weakness:
Op Strength:
Op Weakness:
Other:

Opponent:
Tournament:
Score:
My Strength:
My Weakness:
Op Strength:
Op Weakness:
Other:

Opponent:
Tournament:
Score:
My Strength:
My Weakness:
Op Strength:
Op Weakness:
Other:

NON-TENNIS

Highlights:

Hobby I am most enjoying:

NOTES

MY NUMBERS FOR THE WEEK

Hours on Court:

Hours on Fitness:

Work Ethic / 10:

KEY TAKEAWAYS

Episode Notes:

Chapter Notes:

Topic Notes:

GOALS

What I would like to accomplish and/or improve this week:

What I am working on:

REFLECTION

Favourite part of the week I am most pumped for:

TALK IT OUT

What I want to discuss with my coach this week (and note if I did discuss this and the outcome):

What I want to learn more about this week (and note if I did in fact learn more about this):

NOTES

Top 3 superpower moments:

Any goal updates?

PROGRESS

Top 3 key learnings to bring forward:

Biggest motivator:

3

TRAINING SESSION DIARY

Monday:

Tuesday:

Wednesday:

Thursday:

Friday:

Saturday/Sunday:

TOURNAMENTS

Tournament I'm playing and/or following:

Favourite player to watch in-action:

IMPROVEMENTS:

What I want to improve?

Biggest distraction?

How was my nutrition?

MATCHES/RESULTS

Opponent:

Tournament:

Score:

My Strength:

My Weakness:

Op Strength:

Op Weakness:

Other:

Opponent:

Tournament:

Score:

My Strength:

My Weakness:

Op Strength:

Op Weakness:

Other:

Opponent:

Tournament:

Score:

My Strength:

My Weakness:

Op Strength:

Op Weakness:

Other:

NON-TENNIS

Highlights:

Hobby I am most enjoying:

NOTES

MY NUMBERS FOR THE WEEK

Hours on Court:

Hours on Fitness:

Work Ethic / 10:

My [Tennis] Black Book: Federer, Nadal, Serena, Djokovic, ________

KEY TAKEAWAYS

Episode Notes:

Chapter Notes:

Topic Notes:

GOALS

What I would like to accomplish and/or improve this week:

What I am working on:

REFLECTION

Favourite part of the week I am most pumped for:

TALK IT OUT What I want to discuss with my coach this week (and note if I did discuss this and the outcome):

What I want to learn more about this week (and note if I did in fact learn more about this):

NOTES

PROGRESS

Top 3 superpower moments:

Any goal updates?

Top 3 key learnings to bring forward:

Biggest motivator:

4

TRAINING SESSION DIARY

Monday:

Tuesday:

Wednesday:

Thursday:

Friday:

Saturday/Sunday:

IMPROVEMENTS:

What I want to improve?

Biggest distraction?

How was my nutrition?

MATCHES/RESULTS

Opponent:

Tournament:

Score:

My Strength:

My Weakness:

Op Strength:

Op Weakness:

Other:

Opponent:

Tournament:

Score:

My Strength:

My Weakness:

Op Strength:

Op Weakness:

Other:

Opponent:

Tournament:

Score:

My Strength:

My Weakness:

Op Strength:

Op Weakness:

Other:

TOURNAMENTS

Tournament I'm playing and/or following:

Favourite player to watch in-action:

NON-TENNIS

Highlights:

Hobby I am most enjoying:

NOTES

MY NUMBERS FOR THE WEEK

Hours on Court:

Hours on Fitness:

Work Ethic / 10:

End of Month Block Review & Additional Notes

KEY TAKEAWAYS

TALK IT OUT

GOALS

REFLECTION

TRAINING SESSION DIARY

NON-TENNIS

PROGRESS

IMPROVEMENTS

NOTES

MATCHES/RESULTS

Opponent:

Tournament:

Score:

My Strength:

My Weakness:

Op Strength:

Op Weakness:

Other:

Opponent:

Tournament:

Score:

My Strength:

My Weakness:

Op Strength:

Op Weakness:

Other:

Opponent:

Tournament:

Score:

My Strength:

My Weakness:

Op Strength:

Op Weakness:

Other:

September: Block 9

GOALS:

Week 1 of Block 9 Tasks

Episode to listen to: #54
Book Chapters:
Level 1: *The Secrets to Optimal Coaching Success (2018)* – 1.3
Level 2: *What is Your Game Missing? (2019a)* – Ranking 46 *and* 48
Level 3: *The Science of Elite Performance (2017)* – Chapter 9: Understanding the Athlete – Catering to the Athlete's Learning Needs

Topic: The Power of Song

A powerful tool and equally powerful motivator, the innate power of song is often under-utilised and yet the benefits outweigh the decisional-balance of not to integrate song into practice and/or training sessions for its sheer advantages. The contextualised to habitual behaviours that are individual to you when music is heard is a power in and of itself. Coursing through your veins (not literally, although an increase in blood flow may and/or may not be present) is the sound to the beat of the song on play and your movements to muscle memory can become more heightened – your senses come *alive*. Of course, this is relative to you and personal preferences for that *right* song.

That feeling that is embodied when the music is turned up – or down, describes the underlying reason of your 'why' behind that song – not to be mistaken for all songs and/or music or lyrics, but everyone will have their own selection with their own feeling behind a multitude of songs. How you *feel* is the catalyst in this instance and how this emotional recall can be used to your advantage. For all intents and purposes, these songs should be conducive to your performance opposed to detrimental i.e. uplifting songs are encouraged to be on play in contrast to melodramatic songs.

This forms an element of emotional conditioning that is rarely taken advantage of and is incredibly beneficial. Akin to a memory etched in time, song can be a *key* reminder – a point of recall, whether for muscle memory or an outcome. It is important to note here that this conditioning should be done intermittently opposed to relied upon regularly due to competitive play environments where you need to be able to hear your opponent and the ball, among other things, opposed to the song of choice on loudspeaker. But using song to your advantage can be a powerful weapon and can also heighten your senses when conditioned to perform with and/or without the volume up, or down. In one way, song *can* detract from one of the most important aspects of the game – your auditory senses, yet when used *with* intention song can provide that Level of emotional conditioning. Music that is not too loud to detract from your auditory senses, but loud enough that you can hear it clearly, delivers the best of both worlds.

What song can you recall to calm those nerves, heighten your senses when under pressure, or what song is your *go to* when needing to work up to the moment and get that fire burning – you *will* have a song for that. How about the song that zones you out – a melody that plays like white noise and quiets your surroundings until it is just you and the ball, or the song that reminds you how much *fun* you are having playing your best tennis. These songs are *yours* – found over years of listening to countless songs and will vary from person to player. These are the songs, however, that have a Level of imprinting on your emotions, and it is up to you to press *play* and ignite your senses whilst also helping learn new technical applications. This is how a relationship is born between your performance and song.

As in life, emotional regulation is not always the easiest of tasks but once accomplished and steps are put in place to know *how* to regulate, you are able to better control your own emotions and as such, behaviours. Song is a tool that can be used to help build these steps whilst also offering that point of recall to regulate – whether feeling the pressure that is stopping you from hitting through the ball, or your opponent has the lead and you need to make a *mighty* comeback. Irrespective of the cause, if the feeling is present your emotions are at play and learning how to regulate this response will allow you to better control them during play.

> As with all things, it will depend on how song is utilised in training/practice to build your *response* base and then how song can be leveraged in moments like these.

A separate way to view song may be as your personal notebook to recall when in need – a complementary element that really runs the message home. Whether your game *plan* reinforces the need to control your positioning or to attack your opponent's backhand, there will be a point of recall *when* you were able to do just this – perfectly, and as desired. Using song in this instance is an example of repositioning your thoughts to that moment – regulating your emotions to perform as desired, forming a part of your response base. There is no right or wrong, rather this is a personalised approach at your disposal that you can use to heighten your take on emotional conditioning and your own performance outcomes.

A scenario that may be in play, for example, begins with *you* – having been on court for two hours and you are playing in the midday sun. At this point, you are starting to feel tired with slowed movement and it is one set a piece. For the sake of keeping things simple, for this example a three-set match is in play. You can see that your opponent is suffering just as much as you are, but they are not missing a ball. You have started to make several unforced errors, which contributed to your opponent claiming the second set. They are on a high with the momentum in their favour and you are looking to regulate your emotions and recentre so you can take on the third set with more accuracy, more consistency and take back the momentum so you can score the *win*. And this is when the beat comes on.

That beat in your head that you know all too well. You have practiced these moments over and over again. These are the moments you train those extra hours for and why you have put in the *work* to get to this point. You know what tired feels like and that beat reminds you – this is not tired, this is *living*! You settle in and allow your emotions to follow the rhythm of the song and suddenly, it is starting to feel like early in the first set with fresh legs and your game is playing as expected – smoothly, fluidly and *on* the offensive. That beat settles and your emotions become heightened – that tired feeling has been replaced with a hunger to take it up a Level and see if your opponent can match your Level in the third and go the distance. The beat stays with you as does the hunger – the fire to persevere as you repeat to yourself, this is *living*!

You know what happens next. You know how the set is played out and you know who scores the *win*. How do you know? Because this is one of many scenarios you have learned to emotionally *condition* – to form that response base when the pressure mounts and the game and/or match is no longer going your way. These are the conditions – scenarios you're comfortable with handling simply because you've been there before, and that beat, you know it all too well because it has been with you intermittently during these moments that you do not even need to press *play* – you can *feel* the beat in motion simply because it has become a part of your emotional response that has become conditioned – you have achieved a state of self-regulation under pressure.

Call it an extension of being in 'the zone', the power of song can be used to capture your optimal 'State of Play' so you can learn to *play* it on repeat. Embedded in the 8 Keys is the development of this optimised 'State of Play' but there are more components to these inner workings than meets the eye. One of the most logical benefits of song is its intrinsic motivation but above all, song is a tool now at your disposal to help with your emotional conditioning whether in play, or not. But those who do leverage song have a lot of fun doing so when their performance reaps the rewards.

Week 2 of Block 9 Tasks

Episode to listen to: #8
Book Chapters:
Level 1: *The Secrets to Optimal Performance Success (2016)* – 2.1
Level 2: *What is Your Game Missing? (2019a)* – Ranking 49 *and* 50
Level 3: *The Science of Elite Performance (2017)* – Chapter 10: Relevance *and* The Kinematics of Knee Bend – a Brief Overview

Week 3 of Block 9 Tasks

Episode to listen to: #55
Book Chapters:
Level 1: *The Secrets to Optimal Coaching Success (2018)* – 2.0
Level 2: *What is Your Game Missing? (2019a)* – Ranking 51 *and* 51
Level 3: *The Science of Elite Performance (2017)* – Chapter 10: Sport Similarities for Height of Contact

Topic: Age Relevance

Outside the developmental spectrum of the tennis player, age interactions are often commonplace with 'age' not being seen as a barrier to interact with other children, adolescents, or young adults around *your* age. This 'range' becomes almost obsolete once the earlier years of adulthood hit and when age truly resembles a mere number when considering the breadth of age on the WTA and/or ATP Tours. A near irrelevant factor, a player's age is superseded by their ranking and age itself becomes an indicator of longevity at the top of the game.

At this moment in time, both the WTA and ATP Tours have multiple players inside the Top 30 who are 30 years or older. Away from tennis, 30 years of age is *not* considered by any means an 'older' age range but rather when a person may reach that next *stage* of their career in whatever field that may be. Across the sporting world, there are a multitude of sports where the best players in the world are well over this 30 years of age 'threshold'. Similarly, most sports have an 'off season' that lasts from 2-3 months to *longer* whereas the tennis calendar spans near 11 months of the year is mostly unmatched due to its demands all season long. With rest periods limited to a handful of weeks per season for WTA and ATP Tour players, this is where age can impact a player's performance due to inadequate rest periods.

It should be no surprise then once a player, male or female, enters their 30's that oftentimes performance regressions undoubtedly happen due in part to this lengthy seasonal demand and lack of recovery periods, which is pertinent for players to maintain their respective peak performances. Whilst younger players can persevere through the season with limited rest periods within a calendar year due to their age and physiological advantages, although *not* ideal, the 'older' players who close in on the 30 year threshold do not have the speed of recovery alike their younger counterparts and due to the ongoing seasonal load, the limited recovery periods does take its toll on the human body. It's important to reiterate that due to human physiology – *not* a result of poor recovery, these players at or beyond the threshold will eventually be forced to take lengthier breaks during the season to simply recover. The lack of integrated recovery periods at Tour Level will take a toll on the best of players if *not* made available and why it is important *now* to ensure your schedule integrates these periods to ensure you maintain this cadence in the years to come. This is something incorporated in *your* Black Book. For Tour Level players, even though the small recovery periods available can be optimised, these simply are not enough when you are playing back-to-back seasons. But making the time, to safeguard your body will pay dividends by adding longevity to your career – start integrating these practices *now*. Nonetheless, nothing can exempt even the best of players from the ultimate need to say goodbye.

The likes of **Federer**, **Nadal** and **Serena** were able to maintain incredible *peak* performances well past this threshold whilst **Djokovic** continues to play on. Although not all players have been able to maintain these peaks past their 30's let alone *win* close to 20 or more Grand Slams over the course of their career – this has been courtesy of the 8 Keys *and* these *four* players taking control of their schedules to force recovery periods throughout the season. And boy did they reap the rewards! That kind of control is something special and what needs to be *learned*.

If not for these pre-planned recovery periods, in particular in the later years of all *four* of these careers, it is almost guaranteed that no player could maintain these respective peak performance cycles without the mandated recovery and their accomplishments may very well have been a different story. But they *learned* early – the 8 Keys paid dividends long term. Adequate recovery across the season is a powerful part of any players *planning* and if you aspire to continue to play at the highest Level past the threshold, *learn* from these *four* players and their recovery habits and they will serve you now and, in the years to come.

In direct contrast to these *four*, the significant majority of players that were ranked inside the Top 10 a decade ago are not *all* ranked inside the Top 10 let alone the Top 30 today. That is the nature of age and tennis – the constant demand on the body all year round with limited rest periods. For those that have been able to maintain their *peak*, this is how the 8 Keys have continued to serve them. For those who have regressed outside the threshold, this is a telltale sign of how the 8 Keys have become displaced and age has become a more significant factor due to human physiology – this cannot be stressed enough. Some individuals are more predisposed than others to physiological impacts and the tennis player is no different.

Physiological impediments aside, the 8 Keys *do* in fact help extend a player's ability to perform at their peak for longer which is fundamental to consider. It is the maintenance and upkeep of these Keys that is the deciding factor. There are two players currently inside the Top 30 on the WTA and ATP Tours that are over 30 years of age and are former and/or current Top 10 players *with* Grand Slam *wins* to their name *and* have also been identified as integrating the 8 Keys in their game – currently and/or previously. For a player to maintain a ranking inside the Top 30 they still need, however, at a minimum *four* to *five* Keys in their game. This means that whilst the knowledge of the 8 Keys may be present, from a physiological standpoint they may be unable to maintain the additional Keys but can still maintain an incredibly high Level of play to remain inside the Top 30.

This Level of upkeep inside the Top 30 for *more* than a decade for a player in their 30's comes with a personal preference – the *want* to maintain a rigorous schedule scarcely seen across other sports that demands this Level of elite performance for close to an entire calendar year. This can neither be ignored as often players who reach the age 'threshold' do *not* always make the decision for themselves to step away from the game – their body is the one who calls time. As such, to continuously perform at the highest Level and to *maintain* a ranking inside the Top 30 is an exhaustive feat not every player is privy to even if they've *once* been a Top 10 player with the power of the 8 Keys.

> Ironically, tennis can be deduced to a game of *luck* and timing. If a player is to maintain their Level of play *after* the threshold – luck in not sustaining an injury that limits their progressions, and the timing by which the 8 Keys become *knowledge,* this Level of awareness prepares and protects these players against technically-induced injuries that would otherwise place them at a heightened risk of onset in these later playing years.

When it comes to the dispersion of age across the WTA and ATP Tours, players inside the Top 30 range from their late teenage years through to their mid to late 30's – in some cases an astounding twenty-year age gap. This extensive age range underpins the near irrelevance of age during play on the WTA and ATP Tours, especially once a player has either reached this ranking threshold or has been identified to harness all 8 Keys and is steadily tracking towards this ranking range. And this is when the 'upsets' occur on multiple occasions. In recent years, examples range from Swiatek, Gauff, Alcaraz to Sinner who were *ranked* outside this ranking range when they caused 'upsets' and would later go on to be ranked inside the Top 10 and *win* at least one Grand Slam.

For a contrasting view, **Serena** Williams' last Grand Slam *finals* appearance was in 2019 at 37 years of age at the US Open Championships and Novak **Djokovic**, whilst still active, was 37 years of age at his most recent *finals* appearance at the 2024 Wimbledon Championships. Both players have the *most* Grand Slams, of this generation, to their name – **Djokovic** of *all-time*, on their respective Tours and defeated players a decade or more younger than them on multiple occasions across the duration of their careers. What is important to consider here is that longevity was on both of their sides as were the 8 Keys. Of equal importance, both players were once 'tracking' with upsets to their name. However, over the last decade there is a distinct reason why not all players have the capacity to last the test of time and *win* Grand Slams into their 30's and yet *four* of the best players of all time – also including **Nadal** and **Federer**, did exactly that. The commonality amongst all *four* of these players is the 8 Keys – whether in their younger years or in the final years on Tour, the 8 Keys remained steadfast for those Grand Slam *wins*.

> The 8 Keys do not discriminate and when it comes to age there is no difference – either they are integrated and applied or not. **Djokovic** has maintained them and **Federer** *primarily* did before he didn't, whilst both **Serena** and **Nadal's** Level of maintenance dropped in their final seasons, per their respective rankings, *primarily* due to physiological factors at play.

If one thing is clear, however, the 8 Keys stand the test of time. Irrespective of your current Level of play or ranking – inside the Top 100 or Top 30 on the WTA or ATP Tours, or at your local Club, age interactions have an incredibly important purpose. The lessons you can learn from players both younger and older is immense. At the end of the day, both age and experience-based interactions serve a purpose towards your ultimate development as a player. Whether learning what worked well against an older player, or what did not work very well against a younger player, all these micro-lessons will have a greater impact in the years to come. And if you are already on the WTA or ATP Tours, or when this time comes, these lessons will remain but come in varied forms – irrespective of age.

By keeping a performance-based mindset no matter your age, you are better enabled to set your sights on how to go about executing your game – avoiding the possibility of getting stuck in the 'he's better than me' or 'she's younger than me' loophole that has nothing conducive to offer your game and its overall development. Although should these views develop these are merely a reflection of developed biases formed along the way. As such, now you know the role of age and *play* – its irrelevance and how it sets the stage whether at Club Level or on Tour, that offers a diversified field based on performance and where age really is just a number so be sure to check *your* bias.

Week 4 of Block 9 Tasks

> **Episode** to listen to: #23
> **Book Chapters:**
> **Level 1:** *The Secrets to Optimal Performance Success (2016)* – 4.2
> **Level 2**: *What is Your Game Missing? (2019a)* – Ranking 52 *and* 53
> **Level 3**: *The Science of Elite Performance (2017)* – Chapter 10: Coaching Pedagogy

KEY TAKEAWAYS

Episode Notes:

Chapter Notes:

Topic Notes:

GOALS

What I would like to accomplish and/or improve this week:

What I am working on:

TALK IT OUT

What I want to discuss with my coach this week (and note if I did discuss this and the outcome):

What I want to learn more about this week (and note if I did in fact learn more about this):

REFLECTION

Favourite part of the week I am most pumped for:

NOTES

PROGRESS

Top 3 superpower moments:

Any goal updates?

Top 3 key learnings to bring forward:

Biggest motivator:

1

TRAINING SESSION DIARY

Monday:

Tuesday:

Wednesday:

Thursday:

Friday:

Saturday/Sunday:

TOURNAMENTS

Tournament I'm playing and/or following:

Favourite player to watch in-action:

IMPROVEMENTS:

What I want to improve?

Biggest distraction?

How was my nutrition?

MATCHES/RESULTS

Opponent:

Tournament:

Score:

My Strength:

My Weakness:

Op Strength:

Op Weakness:

Other:

Opponent:

Tournament:

Score:

My Strength:

My Weakness:

Op Strength:

Op Weakness:

Other:

Opponent:

Tournament:

Score:

My Strength:

My Weakness:

Op Strength:

Op Weakness:

Other:

NON-TENNIS

Highlights:

Hobby I am most enjoying:

NOTES

MY NUMBERS FOR THE WEEK

Hours on Court:

Hours on Fitness:

Work Ethic / 10:

KEY TAKEAWAYS

Episode Notes:

Chapter Notes:

Topic Notes:

GOALS

What I would like to accomplish and/or improve this week:

What I am working on:

TALK IT OUT

What I want to discuss with my coach this week (and note if I did discuss this and the outcome):

What I want to learn more about this week (and note if I did in fact learn more about this):

REFLECTION

Favourite part of the week I am most pumped for:

NOTES

PROGRESS

Top 3 superpower moments:

Any goal updates?

Top 3 key learnings to bring forward:

Biggest motivator:

2

TRAINING SESSION DIARY

Monday:

Tuesday:

Wednesday:

Thursday:

Friday:

Saturday/Sunday:

IMPROVEMENTS:

What I want to improve?

Biggest distraction?

How was my nutrition?

MATCHES/RESULTS

Opponent:

Tournament:

Score:

My Strength:

My Weakness:

Op Strength:

Op Weakness:

Other:

Opponent:

Tournament:

Score:

My Strength:

My Weakness:

Op Strength:

Op Weakness:

Other:

Opponent:

Tournament:

Score:

My Strength:

My Weakness:

Op Strength:

Op Weakness:

Other:

TOURNAMENTS

Tournament I'm playing and/or following:

Favourite player to watch in-action:

NON-TENNIS

Highlights:

Hobby I am most enjoying:

NOTES

MY NUMBERS FOR THE WEEK

Hours on Court:

Hours on Fitness:

Work Ethic / 10:

KEY TAKEAWAYS

Episode Notes:

Chapter Notes:

Topic Notes:

GOALS

What I would like to accomplish and/or improve this week:

What I am working on:

REFLECTION

Favourite part of the week I am most pumped for:

TALK IT OUT What I want to discuss with my coach this week (and note if I did discuss this and the outcome):

What I want to learn more about this week (and note if I did in fact learn more about this):

NOTES

PROGRESS

Top 3 superpower moments:

Any goal updates?

Top 3 key learnings to bring forward:

Biggest motivator:

3

TRAINING SESSION DIARY

Monday:

Tuesday:

Wednesday:

Thursday:

Friday:

Saturday/Sunday:

TOURNAMENTS

Tournament I'm playing and/or following:

Favourite player to watch in-action:

IMPROVEMENTS:

What I want to improve?

Biggest distraction?

How was my nutrition?

MATCHES/RESULTS

Opponent:

Tournament:

Score:

My Strength:

My Weakness:

Op Strength:

Op Weakness:

Other:

Opponent:

Tournament:

Score:

My Strength:

My Weakness:

Op Strength:

Op Weakness:

Other:

Opponent:

Tournament:

Score:

My Strength:

My Weakness:

Op Strength:

Op Weakness:

Other:

NON-TENNIS

Highlights:

Hobby I am most enjoying:

NOTES

MY NUMBERS FOR THE WEEK

Hours on Court:

Hours on Fitness:

Work Ethic / 10:

My [Tennis] Black Book: Federer, Nadal, Serena, Djokovic, __________

KEY TAKEAWAYS

Episode Notes:

Chapter Notes:

Topic Notes:

GOALS

What I would like to accomplish and/or improve this week:

What I am working on:

TALK IT OUT

What I want to discuss with my coach this week (and note if I did discuss this and the outcome):

What I want to learn more about this week (and note if I did in fact learn more about this):

REFLECTION

Favourite part of the week I am most pumped for:

NOTES

PROGRESS

Top 3 superpower moments:

Any goal updates?

Top 3 key learnings to bring forward:

Biggest motivator:

4

TRAINING SESSION DIARY

Monday:

Tuesday:

Wednesday:

Thursday:

Friday:

Saturday/Sunday:

TOURNAMENTS

Tournament I'm playing and/or following:

Favourite player to watch in-action:

IMPROVEMENTS:

What I want to improve?

Biggest distraction?

How was my nutrition?

MATCHES/RESULTS

Opponent:

Tournament:

Score:

My Strength:

My Weakness:

Op Strength:

Op Weakness:

Other:

Opponent:

Tournament:

Score:

My Strength:

My Weakness:

Op Strength:

Op Weakness:

Other:

Opponent:

Tournament:

Score:

My Strength:

My Weakness:

Op Strength:

Op Weakness:

Other:

NON-TENNIS

Highlights:

Hobby I am most enjoying:

NOTES

MY NUMBERS FOR THE WEEK

Hours on Court:

Hours on Fitness:

Work Ethic / 10:

End of Month Block Review & Additional Notes

KEY TAKEAWAYS | GOALS

TALK IT OUT | REFLECTION

TRAINING SESSION DIARY | PROGRESS

NON-TENNIS | IMPROVEMENTS

NOTES

MATCHES/RESULTS

Opponent:

Tournament:

Score:

My Strength:

My Weakness:

Op Strength:

Op Weakness:

Other:

Opponent:

Tournament:

Score:

My Strength:

My Weakness:

Op Strength:

Op Weakness:

Other:

Opponent:

Tournament:

Score:

My Strength:

My Weakness:

Op Strength:

Op Weakness:

Other:

October: Block 10

GOALS:

Week 1 of Block 10 Tasks

Episode to listen to: #56
Book Chapters:
Level 1: *The Secrets to Optimal Coaching Success (2018)* – 1.4
Level 2: *What is Your Game Missing? (2019a)* – Ranking 54 *and* 54
Level 3: *The Science of Elite Performance (2017)* – Chapter 10: Learning Theory – Considering Sociological Influences

Topic: Persistence v. Walking Away

You are immersing yourself in *your* Black Book undoubtedly in part due to persisting as a player. Whether you are inside your first decade of play, second – or beyond, there is that driver inside of you that is yet to relent. But there is no guarantee that your current 'State of Play' nor mindset will continue to endure without the right 'guides' in place. These guides serve as a point of reflection and **My [Tennis] Black Book** is one of many forms, but above all is your resounding *why* to persist and maintain *your* truth.

As with all things in *your* Black Book, there is no right or wrong but what resonates true for *you*. There is of course a commonality that exists amongst players and that is an unrelenting desire – a joy, a passion, a *love* to why you continue to play the game. But persistence, at times, will not always seem as clear and that's where walking away may seem like the preferred choice. What is important here is to distinguish between the *why* that serves to guide you, and *your* own inner compass.

Some players will choose to play the game forever and *not* want to walk away from the sport given the decision was theirs to make. But that is not always the case due to the limits of the human body. Similarly, *new* generations will come to play at the elite Level and *new* Top 10 players will make their mark every season whilst when a *new* generation is defined, alike this season, a *new* overhaul will also take place as *new* milestones are made. As such, persistence is no guarantee to *maintain* a Top 10 ranking when a *new* era of play arrives – unless your game keeps pace with the *new* Level of play.

Persistence takes on a different form, behind the scenes, when you ask yourself the question of whether to persevere *or* if enough really is enough. This is where your own resounding *why* serves its greatest purpose. After all, there is no guaranteed timeline at the elite Level of play nor at the developmental, but there is a common trajectory and your *why* is the common theme. But it is far from gloom. There is a guarantee that you *will* have innumerable good days and whilst there will be the intermittent bad, these will be vastly outnumbered by the cherished moments that align with your *why*. Irrespective of your ranking or where you are at on the Pathway to the Long Game, these days will coexist – the good *with* the bad, and *how* you handle these days will depend on the comfort afforded to you by your own inner compass.

Unfortunately, walking away from the game prematurely happens to a sizeable number of players due in part to their why *not* being given the attention it deserves and the lack of support available. But by developing your *why* and knowing where support is available, you will be less susceptible to external noise that may directly influence your persistence to play or make you question whether it is time to walk away *not* on your terms.

> What often differentiates the player that persists to those who opt not to, is *how* they handle the good and the bad days which in turn directly loops back to their *why*. If their why is *not* reflected upon, how a player handles these days will steadily deteriorate and their decision may not wholly be on their terms due to the unbalanced view developed between these good and bad days.

The loop that reinforces a player's *want* to play the game, needs to consider the work that goes into attending to this *why* – embedded in *your* Black Book as a consistent form of reflection. The desire to learn *how* to progress, *how* to improve, or *how* to edge closer towards the Top 10 are some of a number of questions you may ask persistently, *seeking* answers to *how* to reach that next peak performance. Alternatively, you may be more subdued and content with your current Level of play and would simply like to *maintain* your current baseline. Whilst there is no right or wrong, there is a *key* difference between the player more likely to progress towards the Top 10 *and* the 8% in contrast to the player who will happily maintain a ranking outside of the Top 50 but possibly inside the Top 100 if they have at least a *few* Keys to their name.

The real divider between *your* decisions comes down to the Level of support available. If your *why* is attended to and it remains steadfast as your 'guide' but the support is *not* accessible, then the Pathway towards the Top 100, let alone the Top 10, becomes more daunting and *your* inner compass will need to *seek* the tools that can help and contribute towards your decisional-balance. This is when the real battle begins to *know* better and stay true to your *why*. To help *navigate* these periods of doubt or to *know* the next best right move, the real strength comes in putting together your own team to ultimately build the support you need. And complemented with the knowledge of the 8 Keys, your *why* will be stronger than ever.

The ability to persist becomes a lot more streamlined and a natural aspect of your core driving force when you have a team to help motivate, support and guide you – you do *not* need to do all the work yourself. But most players either do not have a team unit *available* or opt to not put in the work to build one of their very own due to the common misconceptions around the advantages, or rather disadvantages, of a team in a single-player sport. Unlike multi-player *team* sports, the team of a tennis player can come in varied shapes and sizes. At the elite Level of play these teams become more fine-tuned and robust that are responsible for keeping the best players in the world inside the 8% but do not be fooled – there is a reason not *all* players inside the Top 10 at the beginning of a season *finish* the season inside the same ranking range. The interrelationships of the team matters – regardless of size.

In the initial stages of developing your team, it is important to remember that there is no right or wrong, it is what or who works best *for* you. The triangular relationship is often a starting point whereby this includes you, your coach and your parent and/or guardian. This can take various forms even if it means you and both of your parents, or you and your best friend plus coach. Parents and/or guardians do not always need to be considered in this triangular relationship, although it is encouraged in your *first* decade of play to have someone who is responsible for your wellbeing, *outside* of the tennis environment, to be kept in the loop. For those of you who are inside your *second* decade of play, your team may consist of, for example, you, your coach, your hitting partner, and/or your physiotherapist or trainer. The right mix is up to *you* and what works for you, but what needs to be kept front and centre is your *why* and for this *why* to be shared with the team and for it to receive their full support. If your *why* is not shared or supported by your team the problem of not having *accessible* support remains and the decisional-balance between your good and bad days remains. It is therefore counterintuitive to have a team *without* the support – they should go hand in hand if you aspire to head towards the Top 10 *or* simply would like to keep playing at your current Level. The people around you – *your* team, matters. The Pathway towards a Top 10 ranking is not for the faint hearted and whilst the good days *will* always outweigh the bad, it is important to keep in mind your *why* to make those bad days that little bit easier. The same applies for the team you have in place and if this is a new concept, it is important to begin with those conversations around your *why* and to ask if they would like to be on your team and support you. These will be relationships built over time and irrespective if you are ranked inside the Top 50 or Top 10, the interrelationships of your team matter *plus* unequivocal support behind your *why*.

Week 2 of Block 10 Tasks

> **Episode** to listen to: #32
> **Book Chapters:**
> **Level 1:** *The Secrets to Optimal Performance Success (2016)* – 5.1
> **Level 2**: *What is Your Game Missing? (2019a)* – Ranking 58 *and* 59
> **Level 3**: *The Science of Elite Performance (2017)* – Chapter 11: Scope for the Movement Characteristics of the Player's Body at Point of Contact in the Technique

Week 3 of Block 10 Tasks

> **Episode** to listen to: #57
> **Book Chapters:**
> **Level 1:** *The Secrets to Optimal Coaching Success (2018)* – 2.1
> **Level 2**: *What is Your Game Missing? (2019a)* – Ranking 62 *and* 63
> **Level 3**: *The Science of Elite Performance (2017)* – Chapter 11: Knee Bend and the Tennis Serve

Topic: Gender Interactions

Whilst there is a long running subjective discussion on Gender Interactions in tennis, the role of physiology *and* biology is often forgotten. Male tennis players are physiologically (and biomechanically) different to female tennis players with greater amounts of testosterone, for example, that can help a player develop *more* speed, endurance and power naturally simply by being born a male. In contrast, female tennis players do *not* have equal amounts of testosterone to enhance their performance but they do have naturally occurring oestrogen in higher amounts by simply being born a female, which does *not* have reciprocal benefits i.e. those born with a Y chromosome have a natural performance enhancer.

From a biomechanical perspective, male tennis players (and in general) have a different build to female tennis players (and in general) that allows the male player an additional advantage, including but not limited to, longer and more robust levers to further optimise the output of their kinetic chain. But should a female player match these biomechanics they have the technical capacity to close in on this range but without the added benefit of testosterone, the female player will naturally fall short providing similar training regimes are followed.

Due to human biology, it is important to reiterate that females are at a performance disadvantage in contrast to males when it comes to energy production i.e. force, power, speed, stamina and *more*. This is not to say, however, that one sex is superior to the other, rather this topic is important to understand *how* the game is played and the maximum capacity of a male will differ to that of a female. By all accounts, each gender has its own respective strengths, but there's also a reason why the game is played differently between genders.

What's paramount to consider, however, is that for a female player to achieve similar feats to a male player their Level of play is in fact *higher* – they've had to work harder to produce equal amounts of power, for example, because they do not have the benefit of a Y chromosome. And that's a fascinating window into the *work* required by both respective genders. In respect to how the game is played, the female player needs to play *smarter* as they do not have the advantage of equal levels of power at their disposal. Whilst the best female players in the world do have power, it is important to make the distinction between developed and naturally occurring power. For example, male players do not need to naturally work as hard to produce equal amounts. For a female player to reach these heights, the work is immense. Now, that's not to say a male's work is not immense, rather a male players consistent power output at equal levels will generally be higher than that of a females, but in instances when a female players is able to produce more power, this shouldn't be confused with naturally produced power. In other words, they've had to do more *work*.

> *Why* are the physiological differences between genders important to consider? The physiological differences between genders are often overlooked at the high performance Level – during and/or post adolescence when the *real* boost kicks into play and by paying attention to these changes allows the necessary adjustments to be made for the player to more optimally perform.

This can cause female players to misperceive their Level of play when in fact they are outperforming the opposite gender by simply keeping pace –

irrespective of the scoreline. But this is not strategic, rather the male player has an all-out advantage courtesy of their hormones. But there is an exception to the rule to *overtake*. If a female player has sounder biomechanics and integrates the 'technical' proponents of the 8 Keys into their game i.e. The V by Dr B ©, then they are at a greater advantage. Of course, if both male and female players integrate the properties of The V by Dr B © at the same rate and frequency, then the Y chromosome wins out.

But not all players have the knowledge of The V by Dr B © nor the 8 Keys. As such, gender *almost* becomes irrelevant. Although should the knowledge share be equal and reciprocated, the natural physiological advantage remains. There are other biological advantages between genders that are not gender-dependent such as height or genetics, but males are predisposed to increased muscle mass to greater lung capacity, which positions them to withstand greater loads on the human body in contrast to the natural capacity of an *elite* female. Irrespective of the Level of *work*, biology will always have its place.

What is interesting to consider, nonetheless, is that these performance advantages do not come between a female's and male's performance until a player's adolescent years. Similarly, not all male tennis players will go through the same changes at the same time due to different periods of growth and the same applies for female tennis players. Before adolescence takes place, gender is *not* indicative of a player's capacity to perform. This is the perfect time to ensure both female *and* male players learn to interact and understand the physiological differences soon to be at play and how a player's game, respectively, needs to be and/or can be modified to account for these changes.

When considering the physiological differences between genders it is simple to understand why Grand Slam play opts for a longer format when male players are in action. This is the equivalent of the physiological advantage being acknowledged and put to *work*. It is important to note, however, that a female player can still be on court for close to three hours whilst a male tennis player will be finishing a four-set match in *less* than that time. After all, tennis is not simply a game of strength and endurance but strategic play and if two players are equally matched in both power and strategy, it is arguably going to be a long-fought battle. And this is why accounting for human physiology and biology is fundamental to understanding the demands of the game.

From a strategic perspective, the best tennis players in the world have a balance between power and strategy and know *how* to leverage these tools. For others, power dominates their game or strategy – but not both. Irrespective if a player is on the WTA or ATP Tours, once they are on the Pathway to progress towards the Top 100 and continue their trajectory through to the Top 50 and ultimately the Top 10 there is an underlying reason they have been able to produce consistent peak Level performances and *yes*, the combination is *key*. The 8 Keys embodies the fusion between both power and strategy and players who harness these tools reap the rewards. To be clear, the data continuously states that there is *no* difference between male and female players when it comes to the mandated technical advantage and what's embodied in the 8 Keys. As such, when the best of both worlds reach a state of autonomy in your game, greatness happens.

There's a reason why **Serena** remains the *most* remarkable woman to play the game. As far as physiological logic is concerned, **Serena** surpassed every milestone before her except for the *all*-time singles record; however, the Level of play delivered by **Serena** was unmatched by previous generations. The fusion of both power and strategy can only arguably be aligned with that of **Federer**, **Nadal** and **Djokovic**, with all *four* players remarkable in their own right. As a female, **Serena** had to *work* to match the power of **Federer**, **Nadal** and/or **Djokovic** without a natural physiological advantage, whist strategically, all *four* players had the edge.

And this is where increased gender interactions can become incredibly rewarding. Whilst both female and male tennis players are encouraged to interact and compete against one another during their developmental through to initial high performance years, this changes during tournaments play and the Level of engagement gradually declines if not *encouraged*. Interestingly, both genders offer compelling advantages for the opposite sex. As the female player doesn't have the natural advantage of power, they need to refine their strategic play. But as the male player naturally has power to more readily develop, their strategic player is not always front and centre. By working together, both male and females can aide in the development of the *other*.

At Tour Level, the gender exchange is not as common as often amateur male tennis players can offer the female player power, force *and* speed during a training session. These amateur players typically are not on tour, or rather at the elite Level, as power they have but their strategic play is not as the Level needed to advance further in the rankings. In contrast, amateur female tennis players who produce ample amounts of power and have developed their strategic play are not present as hitting partners simply due to the fact that this combination is lethal and they are most probably tracking towards the Top 50 or beyond this same season.

Although generalised for accommodation purposes, the differences between female and male players do become exacerbated at the amateur Level of play before closing in again at the higher end of the performance spectrum. Despite the differences in levels of play, this does *not* mean the effort is not equal nor their respective technical capacity, but rather each *sex* has their own constraints imposed by their respective physiology. Irrespective if you are a female or male player, this serves an important purpose in understanding a tennis player's capacity and how your *work* is your own personal benchmark to use with the 8 Keys to reach new heights opposed to gender-based assumptions.

Week 4 of Block 10 Tasks

Episode to listen to: #40
Book Chapters:
Level 1: *The Secrets to Optimal Performance Success (2016)* – 6.2
Level 2: *What is Your Game Missing? (2019a)* – Ranking 66 *and* 67
Level 3: *The Science of Elite Performance (2017)* – Chapter 11: Height of Contact in Relation to Centre of Gravity

MATCHES/RESULTS

Opponent:	Opponent:	Opponent:
Tournament:	Tournament:	Tournament:
Score:	Score:	Score:
My Strength:	My Strength:	My Strength:
My Weakness:	My Weakness:	My Weakness:
Op Strength:	Op Strength:	Op Strength:
Op Weakness:	Op Weakness:	Op Weakness:
Other:	Other:	Other:

KEY TAKEAWAYS

Episode Notes:

Chapter Notes:

Topic Notes:

GOALS

What I would like to accomplish and/or improve this week:

What I am working on:

REFLECTION

Favourite part of the week I am most pumped for:

TALK IT OUT

What I want to discuss with my coach this week (and note if I did discuss this and the outcome):

What I want to learn more about this week (and note if I did in fact learn more about this):

NOTES

PROGRESS

Top 3 superpower moments:

Any goal updates?

Top 3 key learnings to bring forward:

Biggest motivator:

1

TRAINING SESSION DIARY

Monday:

Tuesday:

Wednesday:

Thursday:

Friday:

Saturday/Sunday:

TOURNAMENTS

Tournament I'm playing and/or following:

Favourite player to watch in-action:

IMPROVEMENTS:

What I want to improve?

Biggest distraction?

How was my nutrition?

MATCHES/RESULTS

Opponent:

Tournament:

Score:

My Strength:

My Weakness:

Op Strength:

Op Weakness:

Other:

Opponent:

Tournament:

Score:

My Strength:

My Weakness:

Op Strength:

Op Weakness:

Other:

Opponent:

Tournament:

Score:

My Strength:

My Weakness:

Op Strength:

Op Weakness:

Other:

NON-TENNIS

Highlights:

Hobby I am most enjoying:

NOTES

MY NUMBERS FOR THE WEEK

Hours on Court:

Hours on Fitness:

Work Ethic / 10:

My [Tennis] Black Book: Federer, Nadal, Serena, Djokovic, __________

KEY TAKEAWAYS

Episode Notes:

Chapter Notes:

Topic Notes:

GOALS

What I would like to accomplish and/or improve this week:

What I am working on:

REFLECTION

Favourite part of the week I am most pumped for:

TALK IT OUT

What I want to discuss with my coach this week (and note if I did discuss this and the outcome):

What I want to learn more about this week (and note if I did in fact learn more about this):

NOTES

PROGRESS

Top 3 superpower moments:

Any goal updates?

Top 3 key learnings to bring forward:

Biggest motivator:

2

TRAINING SESSION DIARY

Monday:

Tuesday:

Wednesday:

Thursday:

Friday:

Saturday/Sunday:

TOURNAMENTS

Tournament I'm playing and/or following:

Favourite player to watch in-action:

IMPROVEMENTS:

What I want to improve?

Biggest distraction?

How was my nutrition?

MATCHES/RESULTS

Opponent:

Tournament:

Score:

My Strength:

My Weakness:

Op Strength:

Op Weakness:

Other:

Opponent:

Tournament:

Score:

My Strength:

My Weakness:

Op Strength:

Op Weakness:

Other:

Opponent:

Tournament:

Score:

My Strength:

My Weakness:

Op Strength:

Op Weakness:

Other:

NON-TENNIS

Highlights:

Hobby I am most enjoying:

NOTES

MY NUMBERS FOR THE WEEK

Hours on Court:

Hours on Fitness:

Work Ethic / 10:

KEY TAKEAWAYS

Episode Notes:

Chapter Notes:

Topic Notes:

GOALS

What I would like to accomplish and/or improve this week:

What I am working on:

TALK IT OUT

What I want to discuss with my coach this week (and note if I did discuss this and the outcome):

What I want to learn more about this week (and note if I did in fact learn more about this):

REFLECTION

Favourite part of the week I am most pumped for:

NOTES

PROGRESS

Top 3 superpower moments:

Any goal updates?

Top 3 key learnings to bring forward:

Biggest motivator:

3

TRAINING SESSION DIARY

Monday:

Tuesday:

Wednesday:

Thursday:

Friday:

Saturday/Sunday:

TOURNAMENTS

Tournament I'm playing and/or following:

Favourite player to watch in-action:

IMPROVEMENTS:

What I want to improve?

Biggest distraction?

How was my nutrition?

MATCHES/RESULTS

Opponent:
Tournament:
Score:
My Strength:
My Weakness:
Op Strength:
Op Weakness:
Other:

Opponent:
Tournament:
Score:
My Strength:
My Weakness:
Op Strength:
Op Weakness:
Other:

Opponent:
Tournament:
Score:
My Strength:
My Weakness:
Op Strength:
Op Weakness:
Other:

NON-TENNIS

Highlights:

Hobby I am most enjoying:

NOTES

MY NUMBERS FOR THE WEEK

Hours on Court:

Hours on Fitness:

Work Ethic / 10:

KEY TAKEAWAYS

Episode Notes:

Chapter Notes:

Topic Notes:

GOALS

What I would like to accomplish and/or improve this week:

What I am working on:

TALK IT OUT

What I want to discuss with my coach this week (and note if I did discuss this and the outcome):

What I want to learn more about this week (and note if I did in fact learn more about this):

REFLECTION

Favourite part of the week I am most pumped for:

NOTES

PROGRESS

Top 3 superpower moments:

Any goal updates?

Top 3 key learnings to bring forward:

Biggest motivator:

4

TRAINING SESSION DIARY

Monday:

Tuesday:

Wednesday:

Thursday:

Friday:

Saturday/Sunday:

TOURNAMENTS

Tournament I'm playing and/or following:

Favourite player to watch in-action:

IMPROVEMENTS:

What I want to improve?

Biggest distraction?

How was my nutrition?

MATCHES/RESULTS

Opponent:

Tournament:

Score:

My Strength:

My Weakness:

Op Strength:

Op Weakness:

Other:

Opponent:

Tournament:

Score:

My Strength:

My Weakness:

Op Strength:

Op Weakness:

Other:

Opponent:

Tournament:

Score:

My Strength:

My Weakness:

Op Strength:

Op Weakness:

Other:

NON-TENNIS

Highlights:

Hobby I am most enjoying:

NOTES

MY NUMBERS FOR THE WEEK

Hours on Court:

Hours on Fitness:

Work Ethic / 10:

End of Month Block Review & Additional Notes

KEY TAKEAWAYS

TALK IT OUT

GOALS

REFLECTION

TRAINING SESSION DIARY

NON-TENNIS

PROGRESS

IMPROVEMENTS

NOTES

MATCHES/RESULTS

Opponent:	Opponent:	Opponent:
Tournament:	Tournament:	Tournament:
Score:	Score:	Score:
My Strength:	My Strength:	My Strength:
My Weakness:	My Weakness:	My Weakness:
Op Strength:	Op Strength:	Op Strength:
Op Weakness:	Op Weakness:	Op Weakness:
Other:	Other:	Other:

November: Block 11

GOALS:

Week 1 of Block 11 Tasks

Episodes to listen to: #3 *and* #9
Book Chapters:
Level 1: *The Secrets to Optimal Performance Success (2016)* – 1.3 *and* 2.2
Level 2: *What is Your Game Missing? (2019a)* – Ranking 68 *and* 70
Level 3: *The Science of Elite Performance (2017)* – Chapter 11: Coach-Athlete Communication and the Building of a Coaching Pedagogy

Topic: Systems, Promises, Power and Luck

Progressing towards a Top 10 ranking has previously involved a *lot* of luck. Finding the right coach that knows the Pathway *and* the 8 Keys is incredibly rare, *worldwide*. Then there is the financial demand *plus* wondering *if* you are following the *right* Pathway and *if* the countless promises made to you along the way will remain true. Luck in *knowing* the truth, luck in developing the *key* foundations, to luck in finding the right coach for *you* has never been a guarantee. And if you are lucky enough to reach the Top 10 there's luck in staying for more than *two* seasons to solidify your place inside the 8% and remain. But if you have been following *your* Black Book, luck is no longer singular to a select number of players, but rather a *sprinkle* of your choosing goes together with *your* very own Black Book to take you all the way.

Financial, *luck* or otherwise, the multidimensional sides of *navigating* towards the Top 10 have featured extensively throughout this body of work with rigorous data collated over the last twelve *plus* years that continues to grow. The data has provided answers that have *never* previously been known, allowing players to no longer be reliant on luck with an actual Pathway developed for universal access to *plan* towards a Top 10 ranking. But fear not, whilst available due to different rates of progress and development between players, there's room for all but the 8% will reach a new benchmark this generation and that is a guarantee courtesy of this Level of access.

Nonetheless, the scientific discovery of finding and building the Pathway fills a long-sought void in the tennis system – breaking the overwhelming need for luck by giving *you* the knowledge to truly *navigate* towards the Top 10 alike the greats before. Whilst **Federer**, **Nadal**, **Serena** and **Djokovic** needed a little luck along the way, there is no denying the 8 Keys they came to *know*. And now, you can too. Without the 8 Keys at your disposal there remains the need for a *lot* of luck and, withstanding being placed on the backfoot, before clawing yourself out to know better one day and start anew. But now you have the advantage and what you do, is *yours* to choose.

> The importance of knowing and understanding that this Pathway was built for *you* and to fill a void in the 'system' until now, cannot be understated nor underrated. Whilst various 'systems' of promise exist, not one of these holds the 8 Keys nor are absolute.

The pitfalls of these systems are not in their making but in the promises not delivered when they *were* made. No system ever has had the 8 Keys and only a handful of coaches historically have been privy to them. This is not to be mistaken with a coach who has worked with a player to achieve Grand Slam success, but a coach who has taken a player *without* a Top 10 ranking or Grand Slam success *previously* and has been able to achieve these feats with not one, but multiple players. This Level of success is limited to only a few and why the 8 Keys are *that* exclusive. But a coach who takes a player from near naught – outside the Top 100, and achieves this Level of success, *is* privy to the 8 Keys. The distinction is in the Level and the *work*. A coach who can *replicate* this feat from one player to another, who is yet to achieve the distinctive rate of success, knows *how* to deliver **eight by three**.

What this means for *you* is knowledge in knowing the difference and if a coach is yet to firmly integrate the 8 Keys into their coaching schematic, take caution. If a coach is in the process of learning these inner-workings of the game and new science-based milestones for their player to harness, that is a positive trajectory and better when shared with *you*. For the first time ever, this knowledge gives you the power and puts you in *control* of what coach is best for you. And if promises are to be made, financial costs to be exchanged, the immense knowledge-bank dispersed throughout the *Collection* and in *your* Black Book puts you in the driver's seat of *your* career. But take caution as less than 8% of the Top 100 players' coaches per Tour know. This number is scaled down to 0.8% when the Top 1000 rankings are considered on *one* Tour, or 0.4% across *both*. As such, the likelihood to have a coach that knows the truth *and* the power of the 8 Keys is less than mild. But knowing this is already an advantage to use as you choose.

What is upsetting about this 'knowing' is that too often coaches over-promise and under-deliver due to *not* knowing the 8 Keys or are putting in the work to learn them to help *you* better prepare and progress towards that next peak performance cycle. But with knowledge comes power and that power is now *yours*. But due to the known power-tussle in the coach-athlete relationship whereby the player is viewed as the 'child' and the coach as the 'adult' primarily *before* a player has come to understand the power they hold; this is largely ignored for fear of the truth becoming known. A player is not often seen as an equal to their coach but rather must follow their instructions unquestionably. This is aided by parents and/or guardians who equally perceive the coach as the one *with* the 'power' and their child as the one to be instructed. Of course, instruction is fundamental in the developmental years, but knowing the *signs* of an effective coach-athlete dynamic at all levels of play is paramount. By not knowing these signs you become susceptible and/or predisposed to injury and/or regressive performances.

It is important is to highlight that the power dynamic scarcely shifts until a player is well into their second decade of play when they have the confidence to take the reins and *choose* the right coach for their needs. But this changes now. You know better and you have *access* to the 8 Keys inside the *Collection* that places you in control of choosing a coach who knows or is committed to proactively learning these in contrast to coaching their own way *without* interest in helping you integrate the 8 Keys into your game. Again, the decision is yours to make.

Deciding if a coach is right for you is an important decision. Whether you're *nine* years of age, *nineteen* years of age or *twenty-nine* years of age – it doesn't necessarily get easier, but you get smarter, more confident in knowing *how* it works, and when equipped with the 8 Keys it makes it a lot easier to choose. Getting comfortable with being in control of your own game is *key* and over time you'll learn what kind of coach is a good fit for *you* and provides the right *Keys* in need.

It is important to remember that the coach-athlete relationship is not a one-way street. This relationship is *that* important towards progressing towards a Top 10 ranking that it is embedded in *one* of the 8 Keys all to itself. It's a shared dynamic rich with mutual respect and clear communication – embedded in *another key*, to ensure irrespective of age differences that you are listened to, heard and that this is reciprocated with both sides being receptive. For this to form you need to identify that right mix for *you* and one where the coach acknowledges the power tussle, the need for balance of control, and encourages *you* to make the right decision for *you*.

This means when it comes to promises made, I hear you and *feel* the pain – literally. I've been you and walked in your shoes. It's important to understand that coaching is a profession and as such, it is a business. It's just as important to understand that you as a player are a customer and you're signing up for a service. Although this comes in various forms, the breakdown of the 'coach' is oftentimes important to redefine the coach-athlete relationship and to step away from the personal when making the decision of what coach is *right* for you. But it is important to keep your personal worth centre – *you* deserve honesty and truth. And if you aspire to progress towards a Top 10 ranking then your coach should have a sound awareness of the 8 Keys and if not, be prepared to learn them with you until such time each *key* has reached a state of autonomy in their coaching pedagogy, and you have come to understand them openly and integrate them into your game.

Warning signs will be on display if a coach does not know the 8 Keys. False promises are not acceptable, but honesty and the commitment to *learn* are positive signs of a coach who is in the business for the right reasons – to help *you*. There's nothing wrong with a coach sharing that they are not sure what is required to progress towards a Top 10 ranking, *but* they would be determined and thrilled to figure it out *with* you and try. The combination of honesty, openness and willingness to share that they do not know but would like to, is one of many positive signals to look for.

When considering the right coach for you *if* you are ranked inside the Top 50 but have reached a performance plateau, the 8 Keys are what you *seek* (one of the *seven* **didactic modes**). The data continues to share that players who reach this kind of plateau and do *not* progress are the ones who are *not* privy to several of the Keys let alone *all* eight. If anything, this underscores why it is so important to choose wisely and remain in control otherwise you may very well be following a different Pathway and fall short without the game to break into the Top 30 let alone find yourself as part of the 2% who cross the threshold into the Top 10 each season.

Knowing the difference between the right coach for you and those who are not an optimal fit is a matrix of its own kind and you're responsible for differentiating *what* works for you and *what* works against you. This becomes easier as you enter your second decade of play but doesn't go away. There are countless players on both the WTA and ATP Tours who are held *back* from crossing the threshold every season.

> To *navigate* these setbacks, the more empowered you become by harnessing the 8 Keys, the better. And *your* Black Book is designed to do just that – empower you with the knowledge to *know* better and to know more and ultimately *how* to choose *who* works best for you.

One primary setback faced by the developmental player are the Talent Identification (I.D) programs that exist in close to *all* countries with a *tennis* governing body. These programs are primarily flawed due to their outdated systems *and* lack of knowledge and integration of the 8 Keys whilst remaining without a supportive developmental framework that follows the Pathway. As such, the rate of success of a player who is selected for a Talent I.D program is non-significant – *without* a direct correlation with a Top 10 ranking in contrast to those *not* selected. But there is one advantage of selection into these programs and that is the financial help afforded to those who are accepted; however, this comes with its own warning.

Too often Talent I.D recipients succumb to over-training and/or the onset of injury in their earlier years without the proper attention to mitigate these risks with the technical parameters embedded in the 8 Keys. The combination of overload and incorrect technical development increasingly leads to injury and often halts these players from progressing as expected. Unfortunately, by design these programs do not serve an *individual* players best interests as they are not built to accommodate the *singular* player. Of course, exceptions can and do happen, but if you are selected into one of these programs, it remains fundamental to take caution and follow the rulers of *your* Black Book and those of the *Collection.*

To counter the advantage of funding received by Talent I.D programs and to reduce the associated risks of selection, an alternative option has been built. The *Collection* is an entire framework in itself, built from naught to carry you through both your *first* and *second* decade of play then more. To fund a player's first or let alone second decade of play is immense and yet without a distinct Pathway or access to the 8 Keys, this becomes quite costly and problematic for a player to continue to *play* without the anticipated results. That's why *your* Black Book paves the way. By not taking up selection into a Talent I.D program, this allows you to focus on the 8 Keys whilst also reducing associated Academy and/or Club costs by fine-tuning what you *need* to work and *focus* on, whilst delivering a Pathway that allows you to progressively solidifies the 8 Keys into your game. This leaves *you* in control of your *planning* and a player who opts for an approach that integrates the 8 Keys is more likely to progress towards the Top 10 and take their place in the aforementioned 0.4%.

Knowledge is your greatest strength and overriding power when it comes to *navigate* and *plan* towards the Top 10 – this is where the **data** is at its strongest and *yours* for the taking. By side-stepping Talent I.D programs and/or embracing the two *with* the 8 Keys front and centre, affords not simply an added advantage but a significant financial win. But you can reap the same rewards without Talent I.D programs whilst reducing their associated risks to future-proof your game and ensure you are geared towards longevity at the top of the game.

Ultimately, when it comes to luck *or* a sprinkle, knowledge of the 8 Keys will take you all the way. The matrix of *navigating* towards the Top 10 will remain, but with the right tools in place, drawn from the knowledge-bank of *your* Black Book and complemented by your coach's pedagogy, this becomes streamlined with the utmost of *work* – remember there are no shortcuts. The *Collection* will empower you as *your* Black Book guides you, and as your coach-athlete relationship reaches new heights, luck will remain an afterthought. As for that sprinkle, it has been with you all along *if* you've found your way here and those 3D's remain strong.

Week 2 of Block 11 Tasks

Episodes to listen to: #7 *and* #63
Book Chapters:
Level 1: *The Secrets to Optimal Performance Success (2016)* – 1.7
Level 2: *What is Your Game Missing? (2019a)* – Ranking 71 *and* 72
Level 3: *The Science of Elite Performance (2017)* – Chapter 11: Athlete Perception of his/her Coach

Week 3 of Block 11 Tasks

Episode to listen to: #64
Book Chapters:
Level 1: *The Secrets to Optimal Coaching Success (2018)* – 3.2
Level 2: *What is Your Game Missing? (2019a)* – Ranking 72 *and* 73
Level 3: *The Science of Elite Performance (2017)* – Chapter 11: Self-talk as a Learning Tool

Topic: How Rankings are Not the Whole Truth

The 2024 season has been an anomaly for more than one reason. The Top 10 rankings have predominantly remained out of alignment with 50% of respective player performances *not* correlating with the 'best in world' ranking range. Due to this season's anomaly, more players have crossed the threshold – exceeding the 2% marker that the data has previously and rigorously established. But more players have also regressed outside the Top 10, unable to uphold their Level of play nor maintain their established baseline *or* Level of maintenance which has had a profound impact on the collective rankings. But there are *key* lessons to be learned from this season and one of those primarily is that irrespective of the Top 10 rankings they are not always the whole truth.

The back-to-back events of Indian Wells and Miami highlight *key* differences between Top 10 players *at* the time. Whilst there are multiple back-to-back events across the season, these two events primarily feature the best players *outside* of a Grand Slam and with enough time *passed* after the *first* Grand Slam of the season – the 2024 Australian Open. Players who peaked in January have had enough time to reset to peak once again in March. Other top tier events (i.e. 1000's and/or Masters) this season fell shortly after and/or before the remaining Grand Slams and why the selection of Indian Wells and Miami affords a different glimpse into the alignment between a player's ranking, their performances over a *three* to *four* week period and if they do, or do *not* maintain their peak baseline.

What's interesting about these correlations or lack thereof is that several players who were inside the Top 10 at the time were no longer after the 2024 US Open and the *signs* started early in the season. Whilst not all player rankings are considered, *key* correlations can be deduced from a select number of players and their respective performances at this time *early* in the season, and if they were able to maintain this Level of play – correlate with their ranking or could *not* keep pace with the set benchmark of a Round of 16 result or *further* on both occasions.

> If this benchmark is not achieved, it does not necessarily mean these players are not correlated with their Top 10 ranking, but it does show a separation between *levels* of play and a player's Level of *maintenance* at this point in the season which leaves them more susceptible to regressing outside the threshold later in the season *if* noteworthy results do not follow.

On the ATP Tour, Alcaraz *won* Indian Wells. Sinner *won* Miami. Medvedev showed up in the *finals* and semifinals respectively. **Djokovic** fell in the 3rd Round of Indian Wells and opted out of Miami. Sinner fell to Alcaraz in the semifinals of Indian Wells whilst Alcaraz fell to Dimitrov in the quarterfinals of Miami. Dimitrov made the *finals* in Miami and fell to Medvedev in the Round of 16 at Indian Wells. Zverev made the semifinals in Miami and the quarterfinals at Indian Wells. Both De Minaur and Ruud advanced to the Round of 16 on both occasions. Alcaraz, Sinner, Medvedev, Zverev, Dimitrov, Ruud and De Minaur were tracking.

On the WTA Tour, Swiatek *won* Indian Wells and fell in the Round of 16 at Miami. Sabalenka fell in the Round of 16 at Indian Wells and 3rd round in Miami. Gauff made the semifinals in Indian Wells and fell in the Round of 16 at Miami. Sakkari made the *final* at Indian Wells and fell in the quarterfinals to Rybakina in Miami. Pegula lost in the 2nd Round at Indian Wells and made the quarterfinals in Miami. Rybakina opted out of Indian Wells and made the *finals* of Miami. Collins fell in the 2nd Round to Swiatek at Indian Wells before *winning* Miami. Swiatek, Gauff and Sakkari are tracking whilst Rybakina is void simply due to one event played. Outside of the Top 10, Navarro was the sole player to advance past and/or equal the set Round of 16 benchmark by progressing to the quarterfinals at Indian Wells and the Round of 16 at Miami. Paolini made the Round of 16 at Indian Wells and fell to Navarro in the 3rd Round at Miami – before both players were to be ranked inside the Top 10.

From the beginning of Indian Wells through to the completion of Miami, the ATP Tour rankings only changed slightly with Dimitrov breaking into the Top 10 and De Minaur being pushed outside ($n = 11$). On the WTA Tour, all Top 10 rankings remained with only minor adjustments. Due to several Top 10 players not advancing past the set threshold, the rankings at this point in time did not reflect the data and thus remained incomplete.

Of the players that were tracking in March of this season, Sakkari's Level of maintenance was not to withstand the benchmark of the season and a significant rankings regression followed – outside the Top 20. In contrast, Sabalenka's performance was merely a slight lull as she'd go on to finish her most successful season to date with *dual* Grand Slam *wins*. Zheng did not advance past the set threshold but did go on to claim noteworthy performances in the *final* quarter of the season to correlate inside the 50% – a significant turnaround. Paolini went one better for both to finish inside the Top 5 with Sabalenka, Swiatek and Gauff.

As for Navarro, she continued to go from strength to strength to join the 8% whilst Collins was not able to keep pace in the later part of the season and was pushed back outside the threshold in the final weeks of the season. Zheng, Paolini and Navarro were not only firm barrier breakers for the season but all *three* joined the 8%.

Along with Sakkari, Muchova, Jabeur and Vondrousova regressed outside the Top 10 this season which left a larger opening than normal for more barrier breakers to appear. Whilst this represents a 40% Top 10 anomaly, Pegula was a part of this 50% until the latter half of the season, whilst Ostapenko regressed mid-season to increase the margins but alike Collins, remains within reach. At the time, Swiatek, Sabalenka, Gauff and Rybakina were the primary players to uphold the 8% – reduced to 4% before Paolini, Zheng and Navarro joined them and Pegula started to track. Interestingly, Krejcikova started the season inside the Top 10 but soon after regressed, but her Grand Slam *win* this season and her Grand Slam results, allowed her to progress back inside the threshold in the final week. The final place of the Top 10 has been in and out in the final weeks of the season with no Grand Slam correlation, and whilst Collins was more likely to claim this spot courtesy of her Miami *win*, it went to Kasatkina who has been just beyond the threshold for most of the season.

Seven of ten players on the ATP Tour were tracking at Indian Wells and Miami in contrast to three on the WTA Tour. Of those seven players *all* have remained inside the Top 10. But the barrier breakers this season confirm the power of the data and correlations – De Minaur is the exclusive *first*-time barrier breaker this season whilst Dimitrov crossed the threshold again and Fritz, who started the season inside the Top 10 before slightly regressing, found his way firmly back across the threshold courtesy of his *best* season on tour complemented by his *finalist* result at the 2024 US Open. The margins between both Tours this season, however, were strikingly different in the *first* half of the season before mutual corrections took place. Whilst the WTA Tour had a steeper alignment to make, this indicates a new wave of play that has formed, like that of the ATP Tour in seasons prior. Of significance, both Tours now have a realigned 8% at the close of the 2024 season and both Tours remain with 2% of players inside the Top 10 that do *not* correlate.

Despite the significant anomaly of the first quarter of the season, the results of both Indian Wells and Miami share insights into *how* the season unfolded. De Minaur remained steadfast to earn his place as a barrier breaker and had been flagged to become a barrier breaker *before* the season commenced. Similarly, the likes of Paolini, Navarro and Collins were all flagged *early* in the season and there was almost a 100% success rate if not for Collins' end of year hiatus. In contrast, Rune was noted to regress outside the Top 10 this season *before* the rankings showed and so too was Tsitsipas – underscoring the **power** of the data.

But that's not all. There are several players who have tracked intermittently this season but fallen short of becoming a barrier breaker – for *good* reason. These are the players who are setting themselves up this season to further peak *next* season should they continue their current trajectory. Both Badosa and Shnaider were noted to track when they were both ranked *outside* the Top 80 earlier in the season – finishing the season within reach of the Top 10. Paul is the standalone player just shy of the Top 10 who is poised to become a barrier breaker next season, whilst Draper is close behind followed by Musetti and Fils. The Top 10 on the ATP Tour may experience a shift like that of the WTA Tour during the 2025 season if the respective benchmark cannot be upheld.

If there's one thing for sure, the 2024 rankings found their primary *truth* in the final weeks of the season – *not* before. The WTA Tour remained relatively firm 50% of the time over the course of the season and whilst the 8% does reflect the rankings, there is still a greater susceptibility with more than 2% likely to regress outside the Top 10. Similarly, the ATP Tour found a sounder alignment in the final weeks of the season as De Minaur edged closer to the 8% and Rublev regressed to *almost* fall outside, however, next season undoubtedly will be a test for more than 2% to regress – a pattern shared between Tours.

On the ATP Tour, Tsitsipas is on the brink of re-entering the Top 10 and Paul to make his maiden entry. Rune found form in the later parts of the season to place himself back in contention, and then the aforementioned *three* are inside the Top 20. On the WTA Tour, Collins, Haddad Maia, Ostapenko and Krejcikova all found their way in *and* out of the Top 10 this season, but it would be Kasatkina to finish inside the Top 10 and Krejcikova, courtesy of her Grand Slam feats this season that aligns with the Top 5. These two players, however, are inside the 2% but alike the ATP Tour, there are more players next season that are susceptible to regress, with 50% as close to a guarantee to remain, but *not* all Top 5 players are safe. And that's significant.

Whether one from the Top 5 and another two or three from the remaining Top 10, the same applies on both Tours with the ATP Tour more likely to have more shifts during the 2025 season due to its relative steadiness this season. Similarly, the WTA Tour is not expected to have equal amounts of shifts as this season, but at this rate the 2% is more likely to increase to accommodate lower levels of maintenance. If anything's for certain, the 2025 season will have several stories to unfold, but the bigger question is *how* do these insights impact your game?

> Fluctuations in performance happen but the ability to *maintain* your baseline and be in control of your peak performance cycles matter. When the time comes to play a Top 10 player, take comfort in knowing not all numbers are what they seem, as that very player may *not* be correlating, nor their Level of maintenance be reflected in their ranking. This signals a Point of Inception (PoI) to make your move and take the advantage.

The differential between the 8% to regressing outside the Top 10 can be deduced to a player's Level of maintenance of the 8 Keys. A lower frequency in application of the *key* technical properties, leading with The V by Dr B ©, results in greater susceptibility. In contrast, higher rates of application drastically increase a player's probability of *winning* a Grand Slam and maintaining a secure hold well inside the 8% if not the primary *three*. Undeniably, the likes of **Federer**, **Nadal**, **Serena** and **Djokovic** have applied The V by Dr B © throughout their careers, whilst **Djokovic** continues to do so. Expectedly, rates of application change with time and as such, a player who once had a rate of frequency inside the top percentile will eventually *no* longer produce the same rates of application. This was the case for **Federer**, **Nadal** and **Serena** whilst **Djokovic** began to show signs this season that his rate of frequency is starting to fluctuate.

Not surprisingly, players ranked inside the Top 5 are more likely to maintain their hold inside the Top 10 due to a common denominator – a direct correlation between their ranking and performance. As highlighted through the snapshot of Indian Wells and Miami, these correlations held their truth as *key* markers for the remainder of the season. Whilst some players solidified their place inside the Top 10, others had to *fight* in the final quarter to truly earn their place. Interestingly, the 8% had at least one *key* metric early in the season and reproduced these metrics towards the end of the season – *maintenance* that matters. Swiatek and Sabalenka battled for No. 1 whilst Alcaraz and Sinner also vied for No. 1 with the likes of Gauff edging closer, **Djokovic** and Zverev coming between, before there were *three.*

That's all it takes, one stepping stone at a time before you begin to track alike the *three*. Of course, the Top 10 to the 8% have all tracked and continue to track but the *three* this *season* on both Tours achieved remarkable feats from Grand Slam *wins*, season ending Championships to higher levels of *maintenance* dispersed across the season to earn the points to edge ahead. But those outside are incredibly close behind with the likes of Paolini to Fritz and Navarro to De Minaur having breakthrough seasons with greater levels of maintenance and they are not the only ones. The truth of the rankings then is this – stay the course, solidify the 8 Keys, do the *work* and soon enough you'll be on a collision course with the Top 10 and stand where the likes of Swiatek, Sinner, Alcaraz and Gauff did only a few seasons before. Along with Sabalenka and Zverev, all primary *three* this season on both Tours all started somewhere – far from the Top 10. But with the work and the 8 Keys, each of these players made their dreams a reality. Remember, rankings are not the whole truth, your *wins* will not always show in the numbers, but as you establish higher levels of maintenance, the numbers will come around and your ranking will come to fruition. In your own time and armed with *your* Black Book, this can be your story now, too.

Week 4 of Block 11 Tasks

Episode to listen to: #65
Book Chapters:
Level 1: *The Secrets to Optimal Coaching Success (2018)* – 3.3
Level 2: *What is Your Game Missing? (2019a)* – Ranking 74 *and* 75
Level 3: *The Science of Elite Performance (2017)* – Chapter 11: Teaching Games for Understanding and its Role in Improving Performance

KEY TAKEAWAYS

Episode Notes:

Chapter Notes:

Topic Notes:

GOALS

What I would like to accomplish and/or improve this week:

What I am working on:

REFLECTION

Favourite part of the week I am most pumped for:

TALK IT OUT

What I want to discuss with my coach this week (and note if I did discuss this and the outcome):

What I want to learn more about this week (and note if I did in fact learn more about this):

NOTES

Top 3 superpower moments:

Any goal updates?

PROGRESS

Top 3 key learnings to bring forward:

Biggest motivator:

1

TRAINING SESSION DIARY

Monday:

Tuesday:

Wednesday:

Thursday:

Friday:

Saturday/Sunday:

TOURNAMENTS

Tournament I'm playing and/or following:

Favourite player to watch in-action:

IMPROVEMENTS:

What I want to improve?

Biggest distraction?

How was my nutrition?

MATCHES/RESULTS

Opponent:

Tournament:

Score:

My Strength:

My Weakness:

Op Strength:

Op Weakness:

Other:

Opponent:

Tournament:

Score:

My Strength:

My Weakness:

Op Strength:

Op Weakness:

Other:

Opponent:

Tournament:

Score:

My Strength:

My Weakness:

Op Strength:

Op Weakness:

Other:

NON-TENNIS

Highlights:

Hobby I am most enjoying:

NOTES

MY NUMBERS FOR THE WEEK

Hours on Court:

Hours on Fitness:

Work Ethic / 10:

KEY TAKEAWAYS

Episode Notes:

Chapter Notes:

Topic Notes:

GOALS

What I would like to accomplish and/or improve this week:

What I am working on:

TALK IT OUT

What I want to discuss with my coach this week (and note if I did discuss this and the outcome):

What I want to learn more about this week (and note if I did in fact learn more about this):

REFLECTION

Favourite part of the week I am most pumped for:

NOTES

PROGRESS

Top 3 superpower moments:

Any goal updates?

Top 3 key learnings to bring forward:

Biggest motivator:

2

TRAINING SESSION DIARY

Monday:

Tuesday:

Wednesday:

Thursday:

Friday:

Saturday/Sunday:

IMPROVEMENTS:

What I want to improve?

Biggest distraction?

How was my nutrition?

MATCHES/RESULTS

Opponent:

Tournament:

Score:

My Strength:

My Weakness:

Op Strength:

Op Weakness:

Other:

Opponent:

Tournament:

Score:

My Strength:

My Weakness:

Op Strength:

Op Weakness:

Other:

Opponent:

Tournament:

Score:

My Strength:

My Weakness:

Op Strength:

Op Weakness:

Other:

TOURNAMENTS

Tournament I'm playing and/or following:

Favourite player to watch in-action:

NON-TENNIS

Highlights:

Hobby I am most enjoying:

NOTES

MY NUMBERS FOR THE WEEK

Hours on Court:

Hours on Fitness:

Work Ethic / 10:

KEY TAKEAWAYS

Episode Notes:

Chapter Notes:

Topic Notes:

GOALS

What I would like to accomplish and/or improve this week:

What I am working on:

TALK IT OUT

What I want to discuss with my coach this week (and note if I did discuss this and the outcome):

What I want to learn more about this week (and note if I did in fact learn more about this):

REFLECTION

Favourite part of the week I am most pumped for:

NOTES

PROGRESS

Top 3 superpower moments:

Any goal updates?

Top 3 key learnings to bring forward:

Biggest motivator:

3

TRAINING SESSION DIARY

Monday:

Tuesday:

Wednesday:

Thursday:

Friday:

Saturday/Sunday:

IMPROVEMENTS:

What I want to improve?

Biggest distraction?

How was my nutrition?

MATCHES/RESULTS

Opponent:

Tournament:

Score:

My Strength:

My Weakness:

Op Strength:

Op Weakness:

Other:

Opponent:

Tournament:

Score:

My Strength:

My Weakness:

Op Strength:

Op Weakness:

Other:

Opponent:

Tournament:

Score:

My Strength:

My Weakness:

Op Strength:

Op Weakness:

Other:

TOURNAMENTS

Tournament I'm playing and/or following:

Favourite player to watch in-action:

NON-TENNIS

Highlights:

Hobby I am most enjoying:

NOTES

MY NUMBERS FOR THE WEEK

Hours on Court:

Hours on Fitness:

Work Ethic / 10:

KEY TAKEAWAYS

Episode Notes:

Chapter Notes:

Topic Notes:

GOALS

What I would like to accomplish and/or improve this week:

What I am working on:

TALK IT OUT

What I want to discuss with my coach this week (and note if I did discuss this and the outcome):

What I want to learn more about this week (and note if I did in fact learn more about this):

REFLECTION

Favourite part of the week I am most pumped for:

NOTES

PROGRESS

Top 3 superpower moments:

Any goal updates?

Top 3 key learnings to bring forward:

Biggest motivator:

4

TRAINING SESSION DIARY

Monday:

Tuesday:

Wednesday:

Thursday:

Friday:

Saturday/Sunday:

MY NUMBERS FOR THE WEEK

Hours on Court:

Hours on Fitness:

Work Ethic / 10:

IMPROVEMENTS:

What I want to improve?

Biggest distraction?

How was my nutrition?

MATCHES/RESULTS

Opponent: ______

Tournament: ______

Score: ______

My Strength: ______

My Weakness: ______

Op Strength: ______

Op Weakness: ______

Other: ______

Opponent: ______

Tournament: ______

Score: ______

My Strength: ______

My Weakness: ______

Op Strength: ______

Op Weakness: ______

Other: ______

Opponent: ______

Tournament: ______

Score: ______

My Strength: ______

My Weakness: ______

Op Strength: ______

Op Weakness: ______

Other: ______

TOURNAMENTS

Tournament I'm playing and/or following:

Favourite player to watch in-action:

NON-TENNIS

Highlights:

Hobby I am most enjoying:

NOTES

End of Month Block Review & Additional Notes

KEY TAKEAWAYS

TALK IT OUT

GOALS

REFLECTION

TRAINING SESSION DIARY

NON-TENNIS

PROGRESS

IMPROVEMENTS

NOTES

MATCHES/RESULTS

Opponent:

Tournament:

Score:

My Strength:

My Weakness:

Op Strength:

Op Weakness:

Other:

Opponent:

Tournament:

Score:

My Strength:

My Weakness:

Op Strength:

Op Weakness:

Other:

Opponent:

Tournament:

Score:

My Strength:

My Weakness:

Op Strength:

Op Weakness:

Other:

December: Block 12

GOALS:

Week 1 of Block 12 Tasks

> **Episodes** to listen to: #19 *and* #96
> **Book Chapters:**
> **Level 1:** *The Secrets to Optimal Performance Success (2016)* – 3.3
> **Level 2**: *What is Your Game Missing? (2019a)* – Ranking 77 *and* 79 *and* 85
> **Level 3**: *The Science of Elite Performance (2017)* – Chapter 11: Coach-Athlete Relationship

Topic: From Clay, the Primary Two and the Power of Rest

As the red clay settles and the noise becomes that little bit quieter, the grass beckons for those who have had time for their losses to settle and tackle a new surface with renewed hope of a better performance. Yet the tide is still high for the Champions of the 2024 French Open as their new silverware finds a new place to call home. For whilst there are only two who left triumphant, there are more who should indeed be proud of their accomplishments *and* performances.

Anomalies to environmental factors are all a part of the game *and* the finesse of a player's health. So meticulous is the management at the absolute highest Level that even a slight hiccup can throw a spanner in the works. Colloquial or not, it is a player's Level of awareness of their body through to their optimal performance metrics and if they have been timed to peak at the right moment – or have they peaked too soon or arrived with a half-*empty* tank? Then the unexpected players arrive and upsets are the new name of the game for rarely do *all* Top 16 seeds progress to the Round of 16 at each Grand Slam – there is a place for the anomaly on each occasion.

Then comes *your* time, whether now or over the course of the next decade, that Point of Inception (PoI) *will* be found. Depending on your ranking, perhaps you're at your best this season and yet the Top 10 remains out of reach, or you're the player tracking on the radar – outside the Top 20 but inside the Top 100 and those inroads are nonetheless impressive, but not enough *just* yet. For you, the time has come to make your move and steadily collect. Although should you be inside the 8% then you know the nature of the anomaly and *how* important it is to maintain your baseline and *seek* that next peak performance cycle. All it takes is a single opening – a slight window left ajar for the next barrier breaker to cross the threshold, as the tide will turn when a player's maintenance is no longer upheld, and these are the openings to look for and take the advantage.

Every player has their own PoI based on *how* many Keys are at their disposal and *how* many Keys have reached a state of autonomy. A player inside the 8% has a knowledge-bank rife of the 8 Keys, however, should there be a lapse in maintenance, longevity inside the Top 10 is *not* what their game is ready for. This is the power of the data and understanding the predictive analytics inside the *Collection* – *how* these indictors of progress impact your game and *how* you can use them to your advantage. This is fundamental for cyclical progressions. If a player does not incorporate these *key* data-points then their season could change in a blink of an eye – French Open success *or* otherwise.

Swiatek started the season a little slower than Sabalenka who claimed the *second* Grand Slam of her career. But then Swiatek refocused and the latter of the *first* hard court season *plus* the clay season was primarily dominated by her. Sabalenka did regain her form for the clay season and *almost* overcame Swiatek on not one, but *two* occasions – Rome to Madrid and almost had a third chance in Paris if not for her health. Sabalenka then needed time to recover whilst Swiatek played on – one *with* rest and one *without*. This was a prime marker for Swiatek's season to date which remained immense and at full throttle irrespective of her body needing the time to rest. In contrast, Sabalenka took the time but then Swiatek regrouped after Wimbledon and prepared for the Olympics.

A similar tale unfolded between Alcaraz and Sinner as it was Alcaraz who fell short in the *first* hard court swing whilst Sinner claimed his *maiden* Grand Slam. Alcaraz then found his form but *not* before several earlier defeats than anticipated which marked Alcaraz as *less* likely to progress at the French Open. Shortly before, Alcaraz took the time to regroup and it paid dividends. On the other hand, Sinner played on and was favoured to collect in Paris with short rest period between. Interestingly, the player with the longer rest period was able to reset his peak performance cycle, alike Sabalenka after Paris, and Alcaraz overcome Sinner but not without a *fight*. Both Alcaraz and Sinner then prepared for Wimbledon and yet Medvedev got the *win* over Sinner whilst Alcaraz went *all* the way. Then came the Olympics and it was Sinner to follow in the footsteps of Sabalenka and take the time to rest. On the other hand, Alcaraz continued and played **Djokovic** in a repeat of the 2024 Wimbledon Championships but this time, Alcaraz was the one to fall short. Noteworthy, an 'unsuccessful' season by his own records, up until this point **Djokovic** was without a title and yet Gold was his to claim. Rest can be a powerful weapon.

Insofar as cycles are concerned, Sabalenka took the first before Swiatek took the second and the third. If Wimbledon is considered the fourth cycle, Sabalenka was resting whilst Swiatek fell short, so when it came to the fifth cycle – the Olympics, Sabalenka still rested whilst Swiatek couldn't go all the way, falling to Zheng in the semifinals. So, when it came to the sixth cycle, Sabalenka was rearing to go and Pegula started to *play*. If the US Open is considered the seventh cycle, Sabalenka triumphed, however, Swiatek was back in contention but not at the same Level. The eighth cycle, shortly after the US Open was all Sabalenka with Swiatek out of action and allowed Sabalenka to secure enough points to take the number *one* ranking. Come end of season, it was Gauff's Championship and Zheng came incredibly close. But when Swiatek and Sabalenka were not on their respective *highs*, Paolini made inroads to round out the Top 5.

Then it was Sinner securing the first cycle before both Sinner and Alcaraz were close in the second but arguably Ruud, Hurkacz, Tsitsipas and *more* had their chances. Both third and fourth cycles went to Alcaraz – each with a Grand Slam, and for the fifth cycle it was Alcaraz in the *final* of the Olympics whilst Sinner was out, although it was **Djokovic**'s Gold. The sixth cycle fell to a select few, again, but undoubtedly Fritz was gearing up for more. But the seventh went to Sinner with Alcaraz falling earlier than anticipated, before the eighth cycle was another mix from Zverev's steadiness all season or Fritz's newfound form. Come end of season, it was all Sinner ahead of Alcaraz but to round out the Top 5, Medvedev showed up throughout the season but unlike Zverev or Fritz, couldn't make a Grand Slam *final* this season with the Level of maintenance delivered by the other four remaining slightly ahead.

Noteworthy, between Swiatek and Sabalenka, it was Sabalenka who succumbed to a minor injury during the season and between Alcaraz and Sinner, both had their own minor injury to deal with. Asides from these primary two on the respective Tours, **Djokovic** and De Minaur had their injuries, Rybakina had several health concerns, Pegula had her own injury with a few other players inside the Top 10 dealing with minor setbacks. For Swiatek to be the only player of these *two* to not sustain an injury is a significant factor when considering her load over the course of the season. Of the Top 5 across both Tours, Swiatek, Gauff, Paolini and Fritz are the only players who have remained healthy over the course of the season, or of which it was reported, with Zheng, Zverev and Medvedev joining Sabalenka, Sinner and Alcaraz with minor injury and/or health issues. That's 40% of Top 5 players across combined Tours that remained 'fit to play' this season – *not* an optimal metric.

For players to remain 'fit to play' throughout the course of a season, they are reliant on their core technical metrics to keep their body *safe* whilst cautioning the work that is done integrates rest periods to build towards that next peak performance cycle. In Swiatek's case, it is clear that due to not enough recovery periods, those peak performance periods were unable to be reset. Alcaraz was able to maintain back-to-back peaks but then the duration of this peak cycle demanded rest. Interestingly, Paolini had back-to-back peak cycles and the data reports a higher Level of maintenance upheld in the later end of the season. As for Sinner and Sabalenka, their mid-season rest at the expense of the Olympics worked in their favour whilst Gauff, Fritz and Zverev held steady for most of the season.

Of course, there are additional players to consider but what's fundamental to remember is the power of rest and *how* at the very top of the game, players can continue to suffer if their cycles are not managed. Similarly, players with higher rates of application of the 8 Keys are *less* susceptible to injury but until they have the same Level of frequency across the season, their Level of maintenance will fluctuate. Sabalenka and Sinner benefitted from increased periods of rest whilst Swiatek and Alcaraz bore the brunt of limited rest and were unable to keep pace. But this season marked a new milestone for both Tours.

> In respect to Grand Slams, Alcaraz and Sinner dominated with two apiece. This is the first time the *new* generation on the ATP Tour have secured *all* Grand Slams outside of **Djokovic**, **Nadal** or **Federer**. On the WTA Tour, with Sabalenka claiming two *too*, this marked a new rivalry with Swiatek and Grand Slam hauls – who will collect more, or will Swiatek and Alcaraz maintain their Grand Slam leads?

But **Djokovic** plays on and *still* made a Grand Slam *final* this season and collected Gold. Now a part of the former and *new* generation, the 8 Keys continue to serve **Djokovic** with his Level of play he's been able to uphold. Of the primary two on both Tours, Sabalenka claimed both her Grand Slams on hard whilst Sinner did so, too. Alcaraz was across clay and grass – a scarcely done feat, whilst Swiatek's late peaks across Indian Wells and Miami followed through to the clay – grass wasn't to be. This leaves a lot to be desired insofar as surface correlations are concerned for both Sabalenka and Sinner with a high benchmark set. Swiatek has won on both clay and hard – not yet grass, whilst Alcaraz is the sole player to *win* on all *three*. None of these players, however, have claimed *a* Grand Slam a piece at each of the four Grand Slams but Alcaraz and Swiatek lead, respectively.

Be it clay or peak cycle number six or nine, a player will not make it all the way through without rest, otherwise their Level of maintenance will fall. Upkeep of the 8 Keys is vital to repeat one cycle after the next, but this new generation – each Tour with its primary *two* of the season, are yet to learn the lessons of **Federer**, **Nadal**, **Serena** and **Djokovic**. There's only one way a player can withstand the demands of the Tour and maintain their peak performance cycle whilst continuing to build on one baseline to the next, and that's embodied in The V by Dr B © and the principles to metrics held within each respective key. Which means the answer to the question of who will claim more Grand Slams or maintain their lead, is in who will uphold these rulers and continue to *work* towards that next peak – each cycle of each season.

Week 2 of Block 12 Tasks

Episode to listen to: #24
Book Chapters:
Level 1: *The Secrets to Optimal Performance Success (2016)* – 4.3
Level 2: *What is Your Game Missing? (2019a)* – Ranking 87 *and* 90 *and* 92
Level 3: *The Science of Elite Performance (2017)* – Chapter 11: Summary on Coach Communication and Pedagogy

Topic: A Final Lesson (*beginning*)

Week 3 of Block 12 Tasks

Episodes to listen to: #45 *and* #106
Book Chapters:
Level 1: *The Secrets to Optimal Performance Success (2016)* – 6.7
Level 2: *What is Your Game Missing? (2019a)* – Ranking 97 *and* 99 *and* 125
Level 3: *The Science of Elite Performance (2017) – Conclusion*

Topic: A Final Lesson (*middle*)

Week 4 of Block 12 Tasks

Episodes to listen to: #1 *and* #125
Book Chapters:
Level 1: *The Secrets to Optimal Performance Success (2016)* – 1.1
Level 2: *What is Your Game Missing? (2019a)* – Ranking 239 *and* 240
Level 3: *The Science of Elite Performance (2017) – Introduction*

Topic: A Final Lesson (*end*)

MATCHES/RESULTS

Opponent:	Opponent:	Opponent:
Tournament:	Tournament:	Tournament:
Score:	Score:	Score:
My Strength:	My Strength:	My Strength:
My Weakness:	My Weakness:	My Weakness:
Op Strength:	Op Strength:	Op Strength:
Op Weakness:	Op Weakness:	Op Weakness:
Other:	Other:	Other:

MATCHES/RESULTS

Opponent:	Opponent:	Opponent:
Tournament:	Tournament:	Tournament:
Score:	Score:	Score:
My Strength:	My Strength:	My Strength:
My Weakness:	My Weakness:	My Weakness:
Op Strength:	Op Strength:	Op Strength:
Op Weakness:	Op Weakness:	Op Weakness:
Other:	Other:	Other:

KEY TAKEAWAYS

Episode Notes:

Chapter Notes:

Topic Notes:

GOALS

What I would like to accomplish and/or improve this week:

What I am working on:

REFLECTION

Favourite part of the week I am most pumped for:

NOTES

TALK IT OUT

What I want to discuss with my coach this week (and note if I did discuss this and the outcome):

What I want to learn more about this week (and note if I did in fact learn more about this):

Top 3 superpower moments:

PROGRESS

Any goal updates?

Top 3 key learnings to bring forward:

Biggest motivator:

1

TRAINING SESSION DIARY

Monday:

Tuesday:

Wednesday:

Thursday:

Friday:

Saturday/Sunday:

TOURNAMENTS

Tournament I'm playing and/or following:

Favourite player to watch in-action:

IMPROVEMENTS:

What I want to improve?

Biggest distraction?

How was my nutrition?

MATCHES/RESULTS

Opponent:

Tournament:

Score:

My Strength:

My Weakness:

Op Strength:

Op Weakness:

Other:

Opponent:

Tournament:

Score:

My Strength:

My Weakness:

Op Strength:

Op Weakness:

Other:

Opponent:

Tournament:

Score:

My Strength:

My Weakness:

Op Strength:

Op Weakness:

Other:

NON-TENNIS

Highlights:

Hobby I am most enjoying:

NOTES

MY NUMBERS FOR THE WEEK

Hours on Court:

Hours on Fitness:

Work Ethic / 10:

My [Tennis] Black Book: Federer, Nadal, Serena, Djokovic, __________

KEY TAKEAWAYS

Episode Notes:

Chapter Notes:

Topic Notes:

GOALS

What I would like to accomplish and/or improve this week:

What I am working on:

TALK IT OUT

What I want to discuss with my coach this week (and note if I did discuss this and the outcome):

What I want to learn more about this week (and note if I did in fact learn more about this):

REFLECTION

Favourite part of the week I am most pumped for:

NOTES

PROGRESS

Top 3 superpower moments:

Any goal updates?

Top 3 key learnings to bring forward:

Biggest motivator:

2

TRAINING SESSION DIARY

Monday:

Tuesday:

Wednesday:

Thursday:

Friday:

Saturday/Sunday:

TOURNAMENTS

Tournament I'm playing and/or following:

Favourite player to watch in-action:

IMPROVEMENTS:

What I want to improve?

Biggest distraction?

How was my nutrition?

MATCHES/RESULTS

Opponent: ----------

Tournament: ----------

Score: ----------

My Strength: ----------

My Weakness: ----------

Op Strength: ----------

Op Weakness: ----------

Other: ----------

NON-TENNIS

Highlights:

Hobby I am most enjoying:

Opponent: ----------

Tournament: ----------

Score: ----------

My Strength: ----------

My Weakness: ----------

Op Strength: ----------

Op Weakness: ----------

Other: ----------

Opponent: ----------

Tournament: ----------

Score: ----------

My Strength: ----------

My Weakness: ----------

Op Strength: ----------

Op Weakness: ----------

Other: ----------

MY NUMBERS FOR THE WEEK

Hours on Court:

Hours on Fitness:

Work Ethic / 10:

NOTES

KEY TAKEAWAYS

Episode Notes:

Chapter Notes:

Topic Notes:

GOALS

What I would like to accomplish and/or improve this week:

What I am working on:

REFLECTION

Favourite part of the week I am most pumped for:

NOTES

TALK IT OUT

What I want to discuss with my coach this week (and note if I did discuss this and the outcome):

What I want to learn more about this week (and note if I did in fact learn more about this):

Top 3 superpower moments:

Any goal updates?

PROGRESS

Top 3 key learnings to bring forward:

Biggest motivator:

3

TRAINING SESSION DIARY

Monday:

Tuesday:

Wednesday:

Thursday:

Friday:

Saturday/Sunday:

TOURNAMENTS

Tournament I'm playing and/or following:

Favourite player to watch in-action:

IMPROVEMENTS:

What I want to improve?

Biggest distraction?

How was my nutrition?

MATCHES/RESULTS

Opponent:

Tournament:

Score:

My Strength:

My Weakness:

Op Strength:

Op Weakness:

Other:

Opponent:

Tournament:

Score:

My Strength:

My Weakness:

Op Strength:

Op Weakness:

Other:

Opponent:

Tournament:

Score:

My Strength:

My Weakness:

Op Strength:

Op Weakness:

Other:

NON-TENNIS

Highlights:

Hobby I am most enjoying:

NOTES

MY NUMBERS FOR THE WEEK

Hours on Court:

Hours on Fitness:

Work Ethic / 10:

My [Tennis] Black Book: Federer, Nadal, Serena, Djokovic, __________

KEY TAKEAWAYS

Episode Notes:

Chapter Notes:

Topic Notes:

GOALS

What I would like to accomplish and/or improve this week:

What I am working on:

REFLECTION

Favourite part of the week I am most pumped for:

TALK IT OUT

What I want to discuss with my coach this week (and note if I did discuss this and the outcome):

What I want to learn more about this week (and note if I did in fact learn more about this):

NOTES

PROGRESS

Top 3 superpower moments:

Any goal updates?

Top 3 key learnings to bring forward:

Biggest motivator:

4

TRAINING SESSION DIARY

Monday:

Tuesday:

Wednesday:

Thursday:

Friday:

Saturday/Sunday:

TOURNAMENTS

Tournament I'm playing and/or following:

Favourite player to watch in-action:

IMPROVEMENTS:

What I want to improve?

Biggest distraction?

How was my nutrition?

MATCHES/RESULTS

Opponent:

Tournament:

Score:

My Strength:

My Weakness:

Op Strength:

Op Weakness:

Other:

Opponent:

Tournament:

Score:

My Strength:

My Weakness:

Op Strength:

Op Weakness:

Other:

Opponent:

Tournament:

Score:

My Strength:

My Weakness:

Op Strength:

Op Weakness:

Other:

NON-TENNIS

Highlights:

Hobby I am most enjoying:

NOTES

MY NUMBERS FOR THE WEEK

Hours on Court:

Hours on Fitness:

Work Ethic / 10:

End of Month Block Review & Additional Notes

KEY TAKEAWAYS

TALK IT OUT

GOALS

REFLECTION

TRAINING SESSION DIARY

NON-TENNIS

PROGRESS

IMPROVEMENTS

NOTES

MATCHES/RESULTS

Opponent:	Opponent:	Opponent:
Tournament:	Tournament:	Tournament:
Score:	Score:	Score:
My Strength:	My Strength:	My Strength:
My Weakness:	My Weakness:	My Weakness:
Op Strength:	Op Strength:	Op Strength:
Op Weakness:	Op Weakness:	Op Weakness:
Other:	Other:	Other:

A Final Lesson

Before the season closes and before *your* Black Book *reflects* on the past *forty-eight* weeks and reviews the *key* lessons that have unfolded, there's one more expanded **Topic** to *review* with fundamental lessons to be had, irrespective of your *stage* and/or Level. This ***finale*** affords new insights into *how* the best players in the world are shaped across the duration of a season and *how* their respective *peak* performances are not always aligned with their ranking. For those with the 8 Keys firmly embedded in their game, there is a higher rate of application and a higher rate of maintenance upheld in contrast to those who fall *out* of alignment – even for one **Block** or *two* during the season.

As you have come to *learn* over the course of the year, the 2024 season presented an anomaly – for the barrier breaker *and* the 8% with more than 2% of Top 10 players being susceptible to regressing outside of the *best* 10% of the Top 100 on either the WTA or ATP Tours. During a typical season, *two* players join the Top 10 and *stay* – barrier breakers with longevity courtesy of the 8 Keys. A new phenomenon, however, occurred during the 2024 season as this previously (*scientifically*) established ruler went out the window. More players were dismantled during the 2024 season as more players *felt* the rewards of harnessing all 8 Keys. But a player's ability to maintain the heightened Level of frequency required to remain inside the 8% wavered – *more* than typical. A direct consequence of this misalignment reduced the once steadfast 8% – equal to 80% of the Top 10 on each respective Tour, down to 50% on the WTA Tour and 70% on the ATP Tour. The difference between Tours, however, can be deduced by the **Final 8** of each of the *five* biggest events across the 2024 season – all *four* Grand Slams *plus* the 2024 Paris Olympics.

Up until this point, you have traversed a variety of **Topics**, listened to an assortment of **Episodes** that have formed additional building blocks of your *new* baseline and added to your knowledge bank with *new* **Chapters** covered each week. But *now* is your chance to apply this multifaceted learning approach to understand *why* the 2024 season did *not* go to *plan* for some of the best players in the world and *why* others were able to use this to their advantage and achieve breakthrough and/or *best* seasons. After putting in the *work* to rigorously *plan* your season – each week of each Block, you are better equipped to understand *how* this planning process aides a player's performance whilst also understanding the disadvantages *if* this *plan* is not adhered to and/or followed. And the best players in the world are *not* immune from *planning* or consistently adding to their knowledge bank. Rather, these players have *more* notable gaps in their performance *and* results due to the benchmark for upholding the 8 Keys to *maintain* a ranking inside the 8% *and* progress to the **Final 8** across the season – *twice* if not more, courtesy of **eight by three**.

If you're yet to broach **Level** 2 and the Books across *stages* one, two *and* three, you may not already be familiar with the Round of 16 threshold for Grand Slam *wins* through to achieving *replicated success* – terms penned throughout the *Collection* and detailed in these releases. The Round of 16 relevance, however, has merit in framing the explanatory nature of the **Final 8** – *one* more *win*. Whilst the Round of 16 has been identified with its own *key* correlations of player performances *and* results over the duration of a season to a decade of play *and* a player's rate of success – the **Final 8** can be attributed to explaining the 2024 season anomaly whilst adding further weight to those already anticipated by a Round of 16 result or *further*. Importantly, the **Final 8** has been designed to help you *navigate* the rankings and *plan* when to next peak whilst also knowing *how* to identify the truth of the rankings and *why* a player's performance may *surpass* their current ranking and/or *not* align with their ranking range, at all.

Each of the *four* Grand Slams and the Olympics shed new light on player performances at that point in time during the 2024 season as each synopsis of the last *eight* players standing on either Tour depicts a different angle. Across the Australian Open, French Open, Wimbledon *and* US Open, the player standings (contestants/participants) were as anticipated, however, for Paris 2024 these fell somewhat short due to country and/or player requirements/eligibility. Despite this inconsistency, each one of the *five* events is evaluated with *key* data extrapolated to ultimately frame and provide an *additional* component to the existing predictive analytic model detailed through the *Collection*. The extension of this model ultimately allows you to *track* and better understand *your* performances and for the Top 10 player, this *additional* component of the **Final 8** informs you on *your* current trajectory and whether to stay on course or make a sharp detour. And for those of you who are on your way towards a Top 10 ranking, the **Final 8** affords a lesson for each season and *how* a *progressive* entry – in contrast to a rapid-fire result, into the last *eight* is a surefire sign of knowing your time has truly begun.

Final 8 of the 2024 Season

The Final 8 of each of the *four* Grand Slams *plus* the Paris Olympics this season tells a new story that has yet to be shared, until now. Whilst the Round of 16 has been set as a benchmark throughout the *Collection* for good reason, the Final 8 highlights discrepancies that coexist with known correlations between the best players in the world *and* their results throughout a season. For example, the Final 8 of each Grand Slam *and* Olympics 'should' consist of *all* Top 8 players ranked on the WTA and ATP Tours, respectively. Interestingly, this is rarely the case. But it is important to be clear as to *why* the Round of 16 has been set as an absolute benchmark previously and *why* the Top 8 does *not* necessarily hold equal weight.

As has been previously established throughout the *Collection*, the Round of 16 is a known and notable data point not simply for its 'halfway' mark at Grand Slams nor its milestone in making the *second* week of a Grand Slam, but for the sheer correlations that coexist between a player's ranking *and* their results – this is the round that has previously been identified as *the* absolute metric of significance that depicts a player's Level of consistency *with* their results whilst being poised to become a barrier breaker and/or maintain a Top 10 ranking *if* this data point is a continuation – a minimum threshold. In other words, a Round of 16 result or *more* is what the best players in the world are after to be on track to become a barrier breaker – if they are *not* currently inside the Top 10 *and* stay for more than a season. This feat is another important metric disclosed across the *Collection*. But for players already ranked inside the Top 10 the Round of 16 threshold sets a precedent to achieve this result at least *twice* per season – a generous limit, with distinct known correlations if these players will or will *not* maintain their hold inside the Top 10. And that's a faint nutshell depiction of the rewards these *key* results afford but still an almighty benchmark seldom achieved by *all* Top 8 players in any one season. But the *best* either have and/or do *plus* more.

These *eight* players – or *forty* per Tour this season, has not distinctly been addressed before, affording a *new* lens and a *new* view of *how* players have been tracking across a season. Even more, the Final 8 provides a *new* data point to consider but it is important to be clear that this is *not* a new benchmark nor a *new* threshold, rather the Final 8 at each Grand Slam throughout the 2024 season *plus* the 2024 Paris Olympics, showcases a Level of incongruency *with* the Top 10 at the time and can be used to further explain *why* some Top 10 players have been inconsistent in their results this season, some more unusual than others, and provides answers as to *how* to overcome this for next season. What's *key* to note, however, in addition to these inconsistencies is *how* the Top 8 throughout the 2024 season has changed – *how* the 8% (or 0.8% per *your* perspective) has changed and *how* this became a clear signal of the *new* generation – some already known whilst others introduced themselves this season.

Each of the *four* Grand Slams provides additional data to add to *how* a Top 10 player is performing and/or what player is marked to become a barrier breaker this season and/or those who are starting to make their mark to edge closer towards the elusive threshold. The 2024 season presented an opportune time to collate even more advances – *five* respective Final 8's have been included to showcase the biggest *five* events this season. It is important to consider that whilst these are 'big' events throughout the calendar year, each Grand Slam is weighed as *the* most substantial and whilst the Olympics does not come with points for players to accumulate, they come with a piece of *history* that seldom the best players have been able to achieve on more than one occasion due in part to its *most* daunting pressures and timing.

All *four* Grand Slams will soon be shared with their respective Final 8 *plus* the 2024 Olympics – independently, and *how* the respective players were tracking at that point in time of the season. The initial Grand Slam of the season, however, is considered a PoI that sets the tone for the unfolding season, but as will shortly be shared, more players and their respective levels of play fluctuate at either the *beginning*, *middle* and/or *end* of the season or at *multiple* intervals. Due to various PoI's across the season, whereby not *all* Top 10 players exceed the Round of 16 threshold let alone the Final 8 on *all* occasions, this signals that this *new* generation has yet to firmly establish itself and reach a Level of stability. This marker also highlights that seldom does a player have the capacity to *maintain* their peak performance let alone their *baseline* for the length of the entirety of the season. And only after all *four* Grand Slams and the Olympic Top 8 have been shared, will it become clear *how* these results across the 2024 season can help you better *plan* and *perform* this season and for the ones ahead.

My [Tennis] Black Book is *yours* to harness but there remains a pivotal *lesson* to still be learned. Each of the *five* biggest events across the 2024 season have fundamental *key* takeaways to learn *how* to find *your* PoI during a season and to use it to your advantage. Of course, the Final 8 sets a significant benchmark and yet not *all* Top 8 players on their respective Tours will reoccur, but the rankings may tell another story. As such, the Final 8 tells the story that remains untold and hidden between. Each of the Grand Slams *and* Olympics with their respective deductions will be analysed individually in alignment with the 2024 calendar before a collective review of the cumulative Final 8. As you follow, look to *find* openings that may identify *your* possible PoI as these are the instances each season when *new* players arrive and begin their ascension towards the top of the game.

2024 Australian Open

At the beginning of the 2024 Australian Open (AO), on the WTA Tour Swiatek had a slightly *less* than one thousand point lead over Sabalenka – separating the two best players in the world on the WTA Tour. Without a doubt, both Swiatek and Sabalenka were favoured to progress into the later rounds of the AO with both players finishing on *highs* at the end of the 2023 season, but so too was Gauff coming off her maiden Grand Slam *win*. With Pegula and Jabeur having deep runs at Grand Slams across more recent seasons, perhaps both of these players would make an early mark to their season. Rybakina and Vondrousova *with* Grand Slams to their name were arguably in contention to make the Final 8 and then possibly Sakkari with *all* eight of these players ranked inside the Top 8 prior to the commencement of the *first* Grand Slam of the season. To round out the Top 10, Ostapenko had *just* progressed back inside whilst Krejcikova was bounced out – although the seedings were done *before* this occurred, and Muchova completed the list, however, was sidelined at the time.

To preface the Final 8, *seven* seeded players progressed to the Round of 16 and *four* of these players progressed to the quarterfinals meaning 50% of players to make the Final 8 on the WTA Tour at the beginning of the season were *not* seeded. Of the seeded players, *three* were ranked inside the Top 10. That's a significant anomaly that framed the commencement of the 2024 season. Interestingly, all *three* of these Top 10 seeds were in the bottom half of the draw with the top half vacant of any Top 10 player with *one* seed – Zheng who prior to the Australian Open had made *two* Round of 16 or *later* results at the Grand Slam Level. The *three* other players in the top half included Noskova – ranked 50, Yastremska – ranked 93 and Kalinskaya – ranked 75 whilst Zheng was ranked 15 and seeded 12. Interestingly, Yastremska was a qualifier whilst the AO was the *second* consecutive Grand Slam for Zheng to progress through to the Final 8. The bottom half had just *one* anomaly in Kostyuk – ranked 37, plus Gauff, Krejcikova and the defending Champion, Sabalenka, *all* seeded inside the Top 10.

Of the 50% that were seeded in the Final 8, *three* of the four players progressed, however, Sabalenka had to play Krejcikova so only three of these players *could* progress at best. Gauff did so after a three-set thriller against Kostyuk with the first two sets pushed to tiebreaks before Gauff got the edge in the third. A noteworthy performance for Kostyuk to signal a *key* progression. In the top half, Kalinskaya pushed Zheng to three after taking the first set before Zheng proved why she was the seeded player, whilst Yastremska was impressive against Noskova who was responsible for taking Swiatek out in the 3rd Round. These results set up a significant opportunity for a player to progress to their maiden Grand Slam *final* and for either Gauff or Sabalenka to make back-to-back Grand Slam *finals* after both players faced off for the 2023 US Open title. It was Sabalenka who got the *win* in two tight sets over Gauff, whilst on the top half it was Zheng to progress to her maiden *final* but not without a close encounter with Yastremska – placing her firmly on the radar to watch this season.

The 2024 AO *final* was always going to be heavily in favour of Sabalenka to *defend* her title as it was remarkable for Zheng – outside the Top 10 to progress to the *final* and irrespective of the result, to become the *first* 'new' barrier breaker of the 2024 season. Interestingly, Sabalenka's *win* placed her slightly closer to Swiatek in the rankings – who did *not* have to defend as many points as Sabalenka, but due to Swiatek's *first* week loss, Sabalenka edged that little bit closer in the rankings. But the real story of the Final 8 was that whilst Zheng became a barrier breaker, Gauff progressed to *three* in the rankings whilst Pegula progressed to *four* – irrespective of *not* making the Round of 16 whilst Rybakina, who fell early in a thrilling third-set tiebreaker, was bumped back to *five* in the rankings. As for Jabeur, Sakkari and Vondrousova – who fell in the 1st Round to Yastremska, and Muchova who did not play – all *four* players held on to their Top 10 rankings. Interestingly, players ranked outside the Top 10 had more slight regressions in contrast to the Top 10 and Krejcikova, who had been bumped out of the Top 10 prior to the AO, did *not* progress back inside the Top 10 despite her quarterfinal result – ahead of *every* other Top 10 player asides Sabalenka, Gauff *and* Zheng after the AO. This ranking discrepancy alone highlights *how* at the very beginning of the season the Top 10 rankings were out of alignment with 50% of players highly susceptible to regress outside the Top 10 unless their season promptly gained pace to follow a more positive trajectory.

On the ATP Tour, the rankings *and* results were vastly different to that of the WTA Tour which has *not* always been the case and presented somewhat of an early anomaly. Insofar as rankings were concerned,

Djokovic held more than a *two*-thousand-point lead over Alcaraz, whilst Medvedev trailed Alcaraz by over one thousand points and Sinner was ranked *four* with a little more than one thousand points separating him from Alcaraz. Sinner was the only player of these *four* to *not* have a Grand Slam *win* to his name and all four of these players were favoured to progress to the Final 8 of the AO with their respective results. The remaining Top 10 included Rublev ($n = 5$) who was potentially favoured to surpass the Final 8, De Minaur who became a 'new' barrier breaker the week prior to the AO, Zverev and Tsitsipas – both previously showing promising signs in Australia, whilst Rune and Hurkacz rounded out the Top 10.

To set the stage for the Final 8, the Round of 16 comprised of *thirteen* seeds – *nine* seeded inside the Top 10. This is already an impressive statistic with 81% of players seeded at the Round of 16 stage and ultimately, *all* players who progressed to the Final 8 were seeded with *seven* of these players ranked inside the Top 10 *plus* Fritz who only recently departed from the Top 10 prior to the AO. Noteworthy, 90% of Top 10 players made the Round of 16 with Tsitsipas falling to Fritz and De Minaur to Rublev in five. The *only* Top 10 player absent was Rune who fell in the 2nd Round which also marked his initial regression, per the predetermined predictive analytics, for the 2024 season.

When considering the Final 8 *all* but one match went the way of the higher seed – **Djokovic** in four over Fritz, Sinner in three over Rublev, Medvedev in five over Hurkacz – an absolute blockbuster, whilst Zverev upset Alcaraz in four. The remaining *four* players were all firmly inside the Top 10 but interestingly, whilst 50% of these players had claimed their own Grand Slams – **Djokovic** and Medvedev in contrast to Zverev and Sinner – the *only* player not to have made his Grand Slam *final* debut. Therefore, Sinner's *win* over **Djokovic** in the semifinal was *not* anticipated given **Djokovic**'s record at the AO, however, Sinner had been tracking and per the predictive analytics remained on track to *win* his maiden Grand Slam *if* he was able to surpass **Djokovic**. The other semifinal was between Medvedev and Zverev who were both closely matched – demonstrated by their five set battle, but it was Medvedev who edged out Zverev to claim the *win* and progress to his *sixth* Grand Slam *final.*

Come the *final*, Medvedev was the more experienced player *and* had previously been in multiple Grand Slam *finals,* in contrast to Sinner's *first*, placing him as the favoured player but not necessarily per the data. If you had been following Sinner's trajectory, his time was in fact *now*. But what also needs to be considered is that Medvedev played *two* back-to-back five set matches leading into the *final* whereas Sinner played *three* and *four* sets, respectively. And that's what proved to be a telltale sign when Medvedev was *up* two sets in the *final* before Sinner refused to take a backseat and claw his way back – claiming *all* three following sets to lift his maiden Grand Slam trophy. By far, Sinner's triumph placed him in contention for *more* this season and yet post AO his ranking remained at *four* but within reach of Medvedev who was then only a few hundred points ahead. Interestingly, Alcaraz trailed **Djokovic** by *less* than a thousand points due to **Djokovic** *not* defending his title, Tsitsipas slipped back a few places, and Fritz became a barrier beaker *again* this season whilst Rune who had an early loss – absent from the Round of 16, *progressed* in the rankings, marking a distinct anomaly. With 90% of *all* Top 10 players progressing to the Round of 16 *and* the quarterfinals comprising of *all* players inside the Top 10 asides from Fritz, the only rankings regression to take place was to account for the 10% but Rune did *not* regress. Following a similar pattern to the WTA Tour and Krejcikova's success, it was to be De Minaur who progressed to the Round of 16 only to be bumped out of the Top 10.

The *three* unseeded players that progressed to the Round of 16 were Borges – ranked 69, Kecmanovic – ranked 60, and Cazaux who received a wild card, ranked 122 at the time. Whilst the WTA Tour witnessed *more* players with similar rankings progress through to the Final 8 that was *not* to be the case for the ATP Tour due to the rankings holding *true* at this stage of the Grand Slam *except* for the 2% (or 0.2% per *your* perspective). What is important to highlight, however, is that while the WTA Tour had a 50% misalignment post AO, the ATP Tour had a 10% discrepancy. This placed the ATP Tours Top 8 in a better position to *maintain* their current 8% for the duration of the season in contrast to the WTA Tours Top 8 that signalled *room* for *new* players to arrive. And this was a *key* indicator for the turning of the tide.

Last season, the ATP Tour witnessed Alcaraz solidify his place as the *next* in line behind the likes of **Djokovic**, **Nadal** and **Federer** who had come before, whilst Sinner patiently waited for his time to arrive. Alcaraz was the primary *new* player in the spotlight courtesy of his Grand Slam successes whilst Sinner had to take a backseat until his time arrived – at the AO with his maiden Grand Slam. If anything, Sinner's *win* was a warning for all due to his *less* than rapid-fire results in contrast to Alcaraz, positioning him with an anticipated *slightly* higher Level of maintenance for the season. On a similar front, due to Sabalenka's achievement of *replicated success*, this is something only Swiatek had achieved between the two until the AO and as such, positioned Sabalenka to be in *higher* contention this season and placed the two on a similar trajectory should Sabalenka continue her current peak *and* Swiatek capture her peak shortly after the AO. Therefore, alike the ATP Tours primary *four*, the WTA Tour has its primary *three* with Gauff *just* behind Swiatek and Sabalenka – no other player has managed to close the gap at this stage, although Rybakina is undeniably close.

What is similar across both Tours is the *new* rivalry's formed courtesy of the AO results. Whilst Swiatek did *not* progress to the anticipated threshold, Sabalenka gained ground to signal the two best players in the world were edging closer. And whilst Alcaraz did not progress as far as anticipated, like Swiatek, he did surpass the set threshold, yet Sinner was the player to undeniably gain the most ground. Behind these *four* are the likes of Gauff and Medvedev looking for more Slams to add to their tally this season whilst **Djokovic** stays steadfast in contention. But there is also an opening for new players to make their mark this season, as with those inside the Top 10 to potentially join the *new* primary *four* should they surpass the threshold and reappear in the Final 8 at least *two* more times this season.

What's *key* to learn as a player is that the AO Final 8 highlights clear gaps in play. For the ATP Tour player, the 8 Keys are fundamental to have a remote chance of crossing the threshold this season *if* you have yet to do so at least once before. This Level of maintenance learned and integrated throughout *your* Black Book serves its greatest purpose here – a *new* Level of autonomy must be reached to tackle the current Top 10 on the ATP Tour due to most of these players having the 8 Keys *purposefully* integrated in their game. But this may very well change throughout the season. The bigger opening is clear on the WTA Tour per the Final 8 and the lack of Top 10 players present. Whilst the *minimum* threshold has been shared of *how* many times per season a player should surpass the Round of 16 or at the very least, progress to this benchmark, 70% of the current Top 10 now have significant inroads to make throughout the season in contrast to what typically would be 20% *or* 7% and 2%, respectively, when considering the Top 100 players in the world. Simply, *more* barrier breakers this season will appear *if* this pattern continues which for *you* means these lessons are *key* to take as an almighty signal to make *your* move with more players susceptible to regressing outside the Top 10 this season than a standard season of play.

2024 French Open

Without a doubt the Final 8 of the 2024 French Open (FO) were *not* all anticipated. What you can learn from this is multifaceted in nature but in equal parts outside of *your* control. Rather, there will be times throughout your career where the hand that has been delt will simply come with the punches. It is then *your* role to simply *roll* with them until the time comes and you can come up for air. In tennis terms, facing a Top 10 seed in the 1st Round or let alone the 2nd Round after a qualifying birth or slight regression in *your* results, either way there is a bit of luck that does come into play.

Sure enough, on the WTA Tour this happened to Swiatek at the French Open in the 2nd Round, but she was on the receiving end of a Grand Slam Champion who had stacked up *just* as many Grand Slams in comparison. In other words, with *four* Grand Slams a piece at the time, between opponents in the 2nd Round will *never* make sense – ever. But then again, that is the simple fact of the draw. The same can be said for qualifiers who will play another *qualifier* in the 1st Round – a huge opportunity to notch up career defining points to really give that push. Other players will face off against former Top 10 players who have been sidelined due to an injury and are yet to regain their points simply because they have been out of action – *not* due to poor performances. And so, upsets readily occur on both fronts.

Whilst Swiatek was favoured to defend her 2024 title, the question was evident whether Sabalenka would get to the *finals* to have the two best play for the Championship. Keeping in mind the recent history between these two players over most recent weeks with *two* finals contended and Swiatek *winning* both, there was a Level of uncertainty in the air but also a surety if there was a third, this time Sabalenka would walk away with the *win*. Unfortunately, Sabalenka's health was unable to be sustained and she fell slightly early but *past* the primary Round of 16 threshold. This was a positive result for *how* Sabalenka's season continued to unfold, meanwhile that *third* matchup would need to wait. Swiatek was able to *maintain* her Level of play that was recaptured shortly after the AO and become the player with the *most* Grand Slams to her name that is still *active* on Tour. Of course, V. Williams is yet to call time on her career at the time of writing, but Swiatek is leading the charge for players inside the Top 10 *with* Grand Slams to their name with Sabalenka looking to close the gap.

At the time of the 2024 FO both Gauff *and* Collins were marked to potentially progress late into the second week whilst it could have been Sakkari's time to *finally* make her mark at the Grand Slam Level. Yet seasonal results went out the window with all but Gauff regressing in their results who upheld her Level and made it to the Round of 16 threshold. What is important to note here is how *two* players had regressive performances whilst another was able to remain steady but *not* peak when necessary – *three* Top 10 players. Timing matters and so does the relative peak performance baseline and *your* ability to learn *how* to maintain this when it really counts.

The Final 8 at the 2024 FO comprised of 60% of Top 10 players – an improvement from the 2024 AO. Whilst the likes of Swiatek, Sabalenka and Gauff were anticipated, Rybakina and Vondrousova – both Grand Slam Champions in their own right, found their form post AO to make the Final 8 whilst Jabeur a multi-*finalist* at Grand Slams also placed herself in contention for the tile. The remaining two places went to Paolini – ranked 15, and M. Andreeva – ranked 38, but it was neither players *first* appearance advancing to the threshold this season with both having progressed to their maiden Round of 16 result at the 2024 AO when ranked 31 and 47, respectively. This signalled *two* times a piece for both Paolini and M. Andreeva to advance to the threshold despite their rankings being outside the Top 10. Noteworthy, Top 10 players absent from the Final 8 included Pegula, Sakkari, Zheng and Collins to highlight another anomaly in the rankings.

A similar pattern of performance occurred on the ATP Tour with Ruud, Tsitsipas and Hurkacz – all *three* with impressive performances across the clay season, but not *one* of these players could peak when it mattered and claim their maiden Grand Slam *win*. Zverev on the other hand remained steady and in form, whilst still showing up late in the lead up events – his timing was on point to further peak come the FO and he reaped the rewards. Interestingly, the data did pinpoint Zverev to progress towards the Final 8 at a minimum due to varying factors but his rate of success *across* the past decade really underscored his ability to keep pace and perform when it really mattered. As for Ruud, Tsitsipas and Hurkacz, they were simply *not* able to sustain their respective peaks and despite Ruud and Tsitsipas *almost* coming within striking distance, they simply were unable to peak at new heights, however, Ruud was the one who came undeniably close as he fell just shy of the *final*.

What is important to note here is the *longevity* of Zverev's performance and results. A Top 10 player for a significant part of his career and even *when* he was forced out of play, due to a debilitating injury, he was able to climb back into the Top 10. That is the power of the 8 Keys. It is equally relevant to highlight that Zverev sustained a ranking inside the 8% when **Djokovic**, **Nadal** and **Federer** were *all* in action at the *same* time. And then comes Alcaraz and Sinner – yet Zverev is *still* holding firm inside the 8%. But it is one thing to *maintain* the 8 Keys and another to *dominate* play with its *key* technical parameters. Whilst Zverev does in fact have these technical metrics integrated into his game, the likes of Alcaraz and Sinner have simply also integrated these metrics at a *greater* Level – the core differentiator.

When Alcaraz and Sinner faced off in the semifinals it was Sinner who was more likely to progress due to the *consistency* of his season in contrast to how Alcaraz's season had gone up until then. But both players showed positive signs leading up to this encounter and Alcaraz played some of his best tennis of the season, reminiscent of his Level of play last season. Whilst Sinner remained on track, it was Alcaraz who lifted his Level over *five* with some of the best tennis of the season played between both players. This runs home the rulers behind *maintenance*, peak performances and *how* to sustain these peaks when it really matters. The other semifinal that played out was between Zverev and Ruud and whilst Ruud was favoured to advance to *another* FO *final* courtesy of his lead-up form, it was to be Zverev who maintained his peak and pushed that little further to secure a *second* Grand Slam *final* appearance of his career.

The Final 8 of the 2024 FO comprised of Alcaraz, Zverev, Sinner, Ruud, Tsitsipas, **Djokovic**, De Minaur and Dimitrov – 70% of the Top 10. At the time, De Minaur had bounced out of the Top 10 to 11 whilst all other Final 8 players were ranked inside. The *three* Top 10 players absent from this benchmark were Medvedev, Rublev and Hurkacz, however, Dimitrov upset Hurkacz and De Minaur upset Medvedev in the Round of 16 placing Rublev as the 10% anomaly – a similar pattern from the 2024 AO. Insofar as results were concerned, Sinner advanced over Dimitrov, Ruud over **Djokovic** courtesy of a walkover, Alcaraz over Tsitsipas, and Zverev over De Minaur – all in *three*. When it came to the final, it was not surprising that Zverev remained in contention over *five* thrilling sets, before Alcaraz claimed his *third* Grand Slam, as it was none other than Zverev who defeated Alcaraz at the AO. Although within reach, Zverev's ability to maintain the technical parameters slightly wavered when it mattered.

A similar story unfolded on the WTA Tour. Whilst *all* Top 4 players progressed to the Final 8, the Level of *maintenance* of the technical parameters became evident in all but one encounter – Swiatek dominating against Vondrousova in straight sets. All other matches – Gauff and Jabeur, Paolini and Rybakina, M. Andreeva and Sabalenka were all pushed to *three* with mere points separating those who progressed to the semifinals. What can be learned from these encounters is not simply the firm integration of all 8 Keys but the importance of *maintaining* the *key* technical properties until the final point. A slight lull in application opens the door just enough for your opponent to nudge through – that is all it takes when *two* Top 10 players face off and/or players who have been *tracking* consistently over the course of the season.

Both Swiatek and Gauff were anticipated to reach the last *four* of the FO. Again, 50% of the remaining *four* were ranked inside the Top 10 whilst the other 50% – Paolini and M. Andreeva were not; although, they were responsible for their respective quarterfinal upsets over Grand Slam Champions. What is important to note, however, is that whilst neither Paolini nor M. Andreeva were anticipated to progress to the semifinals, both players maintained their baseline and formed their next peak baseline by progressing past the Round of 16 threshold and past the Final 8 benchmark. If either player fell in the quarterfinals this metric would still remain true due to their consecutive peaks. When playing one another in their respective semifinal, however, it came down to *experience* and it was Paolini that progressed to her maiden Grand Slam final. Interestingly, in the semifinal between Swiatek and Gauff, it too came down to *experience* with Gauff simply unable to match the Level of play delivered by Swiatek. So, when it came to the *final* it was no surprise that experience played a factor as the FO had become Swiatek's hunting ground and if anything, it was an absolute thrill to witness the arrival of a new clay queen by claiming her *third* title in a row and *fourth* overall – overtaking **Serena's** *three* crowns.

The learnings across both the WTA and ATP Tours from the 2024 FO included the impact of *anticipated* performances, the Level of play delivered, and the role of *experience* and how these pillars feed back into the integration and ultimate development of *all* 8 Keys. Intertwined and dispersed between the Final 8 was *how* this came to fruition as whilst some players simply *maintained* their Level of play until their opponent lifted, others peaked at the *right* time and were able to demand more of themselves and notch the *best wins* of their career. Be it at *your* Academy and/or *your* first Grand Slam qualifying round, the Final 8 will not always reflect the rankings because players like *you* have the capacity to disrupt the Top 10 and their respective *Level* of play, should it not exceed the benchmark. This is when you can begin to make *your* mark and there are plenty of opportunities over the course of the season. That is all it takes – one progressive *win* at a time and to *find* that window. Should you maintain *your* commitment to developing and cementing the 8 Keys into *your* game, you are more likely to become a player with a Level of play that can be upheld in times of need and collect those career defining *wins* to join those at the top of the game.

2024 Wimbledon

A wild coincidence that despite Wimbledon's Final 8 not being synonymous with current rankings *and* performance projections, the percentages closely resonate with the 8% and the 2% that break through each season. Whilst 75% of WTA Tour players were seeded and the remaining 25% were 'surprise-packets', only 25% of these players collectively held a ranking inside the Top 10. On that basis, 75% of these players were *not* anticipated to progress to the Final 8. Although this is *common* this season, it does highlight the continued misalignment between the Top 10 at the biggest *four* events across the season. Similarly, the ATP Tour saw a drop in Top 10 players progressing to the Final 8 with 50% of Top 10 players progressing yet all *eight* were seeded in contrast to the WTA Tour with *two* newcomers unseeded.

At this stage of the season, the misalignment between Grand Slam results and Top 10 rankings has become more profound. Whilst only a handful of these players have crossed the Round of 16 threshold *twice* across the now *three* Grand Slams of the season, this is cause for concern when considering the true rankings and validity of the Top 10 and the Final 8. But the Round of 16 does allow a higher margin for error and as such does include more Top 10 players across *both* Tours. Nonetheless, the Final 8 is a *high* data point and as promised, tells an intriguing story. These progressions and/or regressions of the Final 8 share an enviable Level of *maintenance* and/or a Level of play this season that does not merit a Top 10 ranking *when* playing a Grand Slam if it cannot be sustained.

Of course, the Final 8 is only one benchmark when a Grand Slam *win* is *key*. But of the *four* chances per season, *two* should at the very least surpass the Round of 16 threshold. Whilst this is a significantly high metric, they are embedded throughout the *Collection* for the simple reason that they really are possible. When considering the likes of **Federer**, **Nadal**, **Serena** and **Djokovic** – irrespective if *still* active or during their careers, these data-points *were* normalised. Likewise, several players across the span of the past decade on both WTA and ATP Tours have consistently been ranked inside the Top 10 and have achieved these metrics. But we are now in a different era and if this new generation is to keep pace, this threshold needs to be met. As a bonus, *now* is the time for you to make your move during these seasons of opportunity and why several unseeded players have already done so and for the likes of Paolini, come within a whisper of becoming a Grand Slam *Champion*.

On the WTA Tour, Rybakina is the only player inside the Final 8 who has remained a Top 10 player *all* season. Krejcikova was inside the Top 10 earlier in the season, however, has since regressed outside. Paolini is the barrier breaker after her *finalist* result at the 2024 FO whilst Svitolina, a former Top 10 player, made the semifinals last season, and Navarro is on track to become a barrier breaker *this* season, whilst Ostapenko sits just outside the Top 10. The *two* unseeded players include Vekic – ranked 49, and Sun – ranked 123 prior to the Championships. Of all *eight* players, Rybakina, Ostapenko and Krejcikova have Grand Slams to their name – accounting for 38% of remaining players whilst 25% of players are inside the Top 10. Whilst this is a regressive appearance of the Top 10 in the Final 8 this season, it is Rybakina, Krejcikova and Paolini's *second* Final 8 result this season that underscores *three* players, irrespective of Top 10 ranking or not, align with this ranking range.

Despite Svitolina not being ranked inside the Top 10 it is *not* surprising that she has reached the Final 8 with her Level of play primarily aligned with a Top 10 ranking. It is surprising, however, that Sun progressed to the Final 8 – her appearance drawing similarities with the AO and those that were *not* expected to progress. Whilst Vekic is also a surprise, her results this season have been steadily tracking and she's previously achieved breakthrough results, although hadn't previously reached a Grand Slam semifinal. As for Krejcikova, after her Level dropped post AO, her Level of *maintenance* at the Grand Slam Level cannot be denied – recapturing her form in time to achieve another Final 8 result whilst Ostapenko held true to form, having remained on the cusp of the Top 10 for most of the season

Although both Paolini and Navarro were not necessarily anticipated per their rankings to reach the Final 8, Navarro had been marked as a potential barrier breaker this season and Paolini had recently secured her place inside the Top 10. These results further reaffirmed the data that *both* players are on track to **finish** the season ranked inside the Top 10. What is interesting to note is that despite Rybakina's ranking, it was Paolini who was primed to potentially claim her maiden Grand Slam even though both Rybakina and Paolini made the Final 8 at the FO mere weeks prior, Paolini was the player to advance to the *finals*.

As far as the results are concerned, Rybakina advanced over Svitolina in the quarterfinals, as did Paolini over Navarro – highlighting *why* both these players have advanced again to this stage of a Grand Slam but why Paolini was the *first* of the two to become a barrier breaker. The results also included Krejcikova defeating Ostapenko which was a surprise per rankings at the time, but both players had ranked inside the Top 10 this season. The remaining result was between Vekic and Sun with *experience* playing a factor in Vekic's *win*. As for the semifinals, 75% of players were seeded and 50% held a Top 10 ranking but only *one* would prevail. It was Vekic who came within mere points of progressing to her first Grand Slam *final*, but it was to be Paolini who held on to secure her *second* consecutive Grand Slam *final* appearance. Between Krejcikova and Rybakina, it was the former Champion who pushed Krejcikova all the way, but it was Krejcikova who managed to hold her nerve and advance to the *second* Grand Slam singles *final* of her career.

The *final* was a rollercoaster. Paolini let nerves get the best of her in the first whilst Krejcikova let her *experience* in Grand Slam *finals* carry her through. Noteworthy is that Krejcikova has multiple Grand Slam *doubles* titles to her name, however, won her maiden singles Grand Slam *outside* the timestamp to achieve 'replicated success'. Interestingly, considering the strength of the data dispersed throughout the *Collection*, Krejcikova is the *first* anomaly since the 'rulers' behind 'replicated success' became known i.e. the likelihood of a player and their capacity to *win* a *second* Grand Slam or more per the predictive analytics. But this season has comprised of multiple anomalies caused primarily by the juxtaposition of the rankings. That said, although Krejcikova was not achieving Grand Slam singles *wins* she was achieving Grand Slam doubles *wins* and this is an area the data do not dissect due to clear 'rulers' in place. This may very well mean in future investigations that there is in fact a marker that exists that strengthens a player's probability of achieving 'replicated success' if they have achieved doubles success across a specific timeline, but that is something that *cannot* yet be stated. If not for the *experience* of Krejcikova, Paolini would have been favoured to *win* her maiden Grand Slam courtesy of her season's trajectory. But this is the **power** of *experience* – a component embedded in the 8 Keys. As such, the 2024 Wimbledon *final* was one of the *best* in recent years and Krejcikova's *win* over Paolini was anything but one sided filled with *key* takeaways for closer inspection.

On the ATP Tour, 50% of Top 10 players advanced to the Final 8 whilst *two* quarterfinals were between *two* Top 10 players. As such, there was never going to be a 100% success rate for Top 10 players to progress through to the semifinals. Most surprising was Sinner *falling* at this stage of the Championships to Medvedev, however, he has been showing up *all* season and it was not that shocking that he came through with the *win* in not three but no less than *five*. Of all the Final 8, Medvedev, De Minaur and Fritz have progressed to the Final 8 now twice this season, whilst **Djokovic**, Sinner and Alcaraz have done so for all *three* Grand Slams. At the 2024 Wimbledon Championships, *two* players presented a slight anomaly – Paul and Musetti, however *both* players were seeded and Paul has been tracking for most of the season.

Between Alcaraz and Paul – primed to become a barrier breaker this season and *why* his progression to the Final 8 had been anticipated, Paul pushed the defending Champion to *four*. Between Fritz and Musetti who peaked at the FO with an 'almost' *win* over **Djokovic**, Musetti came away in *five*. Despite the result for Fritz, he has been showing up *all* season and continues to go from strength to strength after regressing outside the Top 10 earlier in the season per the predictive analytics. Nevertheless, at this rate and at this Level, Fritz is on a positive trajectory to rejoin the Top 10 this season and make a deep run at his home Grand Slam later in the season. As for Musetti, his Level of play is not surprising but his Level of *maintenance* on display is – a signal for more to come. The last of the Final 8 was De Minaur and **Djokovic**, however, **Djokovic** received a walk-over.

The semifinals comprised of Alcaraz advancing over Medvedev and **Djokovic** over Musetti to set up a repeat of last year's *final*. What was most telling about the 2024 Championships was the way Alcaraz achieved *back-to-back* Grand Slam *wins* – from claiming the 2024 French Open then backing it up with the 2024 Wimbledon Championships – something Swiatek unable to do. However, it was Swiatek who impressed leading up to the FO and Alcaraz who came in arguably underdone yet reaped the rewards. As such, Alcaraz has played *less* tennis this season by way of contrast but capitalised on his greater rest periods. As for the *final*, the way the *win* occurred – in straight sets over **Djokovic**, was almost unthinkable and yet Alcaraz's Level of play was on full display. No longer knocking on the door of the next generation, Alcaraz is leading the charge with his *current* Grand Slam haul.

If there is anything that is clear from the 2024 Championships, it is the cumulative rankings displacement across both Tours. The WTA Tour had 2% with the remaining 8% *not* progressing – a reversal of what 'should' have been. The ATP Tour did have 5% of its 10% present, yet it was the *lowest* rate so far this season. As to *why* the immense gap *inside* the Top 10 continues to exist at the Grand Slam Level, the answer is a result of performance correlations – a player's ability to *maintain* their peak baseline, whilst cementing the 8 Keys into their game with limited margin for error. Those with higher margins and/or lower levels of *maintenance* are the ones who are more readily displaced and fall at the initial hurdles, as their games become compromised, and an opening awaits. But there is an additional factor to contend with which is overload, something Swiatek has started to show signs of succumbing to this season in contrast to Sabalenka *not* playing Wimbledon. But both Sinner and Alcaraz did make the Final 8 yet Zverev did *not* – but made the FO *final*, while Paolini surpassed the Final 8 on both occasions and advanced to the *final* on both occasions. The real question then comes down to a player's load and its management. For players who can transition swiftly and smoothly whilst taking adequate rest, a balance is typically achieved, but for those who struggle to balance the *highs* of their achievements *plus* periods of rest, their performances will soon become susceptible to those with the *fresher* legs.

2024 Paris Olympics

To better understand the historical nature of Olympic results and concurrent performances, listen to the respective Special Editions for Paris 2024 on the Beyond Top 10 Tennis Podcast that discusses this in a broader context in alignment with the respective results.

No one really expected the Paris 2024 Olympics to unfold as they did. But honestly, if you have been following and have a grasp on the *known* technical metrics that underscore *how* the data informs these outcomes, you are equally *not* that surprised. But there are some seriously interesting outcomes that are worthy of discussion and *how* these lessons can be applied in *your* game. The biggest is that irrespective if you are having a lengthy lull in your seasonal performances, that does *not* mean a Gold medal is off the table – or your own respective peak performance.

If that is not something that perplexes the *best* in the world, nothing will. The irony here is that despite the accolades that come with Grand Slam successes, the pinnacle of the 'sporting world' is largely denoted by an Olympic medal –

Gold, Silver or Bronze are as close to equal weight of Championship *wins* in other sports, although this view shifts slightly across the tennis landscape. Why? Arguably because only *one* player in the Open Era has won the *Golden* Slam – Steff Graf. No other player has claimed *all* Grand Slams inside a season *plus* an Olympic Gold, including the likes of **Federer, Nadal, Serena** and **Djokovic**.

Interestingly, a Gold on the women's Tour is almost taboo. In the past decade, not *one* player with *a* Grand Slam to their name has claimed Gold. Silver and Bronze are a different story and yet these players are seldom referenced ahead of the Gold medallists. Prior to the last decade, the 2012 London Olympics was quite different with *all* men's and women's medallists having *a* Grand Slam to their name – *big* news. Whilst this has happened intermittently over the years, scarcely have both women's *and* men's medallists had a Grand Slam to their name. But since London 2012 the tide has unfavourably changed.

First, this is a snapshot of the most recent decade of play and then some – using the 2012 London Olympics as a benchmark. And since this point of inception, not *one* player who has *won* Gold on the women's side has been anticipated to claim *the* elusive medal. The men's side is in stark contrast with one of the best players of that season triumphing. From Murray – at the time privy to the 'Top 4' in his own right, going back-to-back whilst remaining the *only* player to defend their Gold in the Open Era; to Zverev in Tokyo 2020 (*held in 2021*) who went on to make his maiden Grand Slam *final* debut and had been ranked inside the Top 10 already for a number of seasons, before Paris 2024 was to be the *first* 'title' **Djokovic** was to claim this season. **Djokovic**'s Gold was almost unanticipated due to his comparative form being unlike seasons before i.e. *less* successful than his typical benchmark, but *still* surpassed the set threshold at each Grand Slam prior – a feat equal to that of only Alcaraz and Sinner. What is important here is that several Top 10 players did *not* play the 2024 Olympics, including Sinner, and despite **Djokovic** facing **Nadal** in the 2nd Round – who was *not* in peak form, he did *not* play a Top 10 player until the *final* who was non other than Alcaraz. A reversal of the 2024 Wimbledon Championships mere weeks prior, the Gold medal match was unquestionably one of the *best* tennis matches played in recent years. These cumulative results across the last *three* Olympics, however, are by no means surprising when considering the more recent *two* across the women's Tour.

The 2012 Olympics witnessed the *best* in the world on display from **Serena** *winning* her *singles* Gold plus *both* Sharapova and Azarenka *winning* Silver and

Bronze, respectively. Interestingly, it was Azarenka who was ranked No. 1 at the time. Like the men, *all* women to achieve a medal at London 2012 were Grand Slam Champions. That was the last Olympics where *all* medallists were to be inside the Top 10 *and* with Grand Slams to their name. In contrast, those who went on to *win* Gold at the next *two* Olympics were primarily *less* than anticipated. The 2016 Olympics saw Puig claim Gold – ranked *outside* the Top 10, whilst Kerber and Kvitova secured Silver and Bronze, respectively and both *with* Grand Slams to their name. Puig had *not* progressed past the Round of 16 at a Grand Slam prior and would go on to achieve this result *once* throughout her career. All other Grand Slam results from Puig were primarily 1st or 2nd Round results. Interestingly, Puig did not rank inside the Top 20 throughout her career and *why* come Rio 2016 Puig was *not* favoured to *win* Gold. Similarly, at Tokyo 2020 (*held in 2021*) the Gold went to Bencic who had previously held a Top 10 ranking prior to the Olympics but had since slightly regressed – *not* favoured by ranking to *win* Gold. As far as Grand Slam results were concerned, at the time Bencic had a quarterfinal *and* semifinal to her name at the Grand Slam Level but had *not* progressed past the 3rd Round during the 2020 and 2021 seasons – *prior* to the Games. By deduction, Bencic was *not* favoured ahead of Svitolina – who won Bronze. Interestingly, it was Svitolina who had *four* quarterfinals and *two* semifinals to her name at the Grand Slam Level. Both Svitolina and Bencic were ranked close to the Top 10 – Svitolina inside the Top 10 whilst Bencic was slightly outside this *key* ranking range followed by Vondrousova, who went on to *win* Silver, ranked inside the Top 50 at the time of the Olympics but had broken into the Top 20 *two* seasons prior courtesy of a Grand Slam *finalist* result at the 2019 FO. Rankings aside, Vondrousova was the favoured player of the *three* with *two* Round of 16 results at the Grand Slam Level leading up to the Olympics – surpassing the results of *both* Bencic and Svitolina.

Come Paris 2024 the *regressive* pattern on the WTA Tour improved as far as the Top 10 rankings were concerned whilst the ATP Tour achieved a greater correlation with the *best* in the world, however, the unanticipated player *remained.* On paper, prior to the 2024 season **Djokovic** was the favoured player to claim Olympic Gold, but both Alcaraz and Sinner had proved to be the better players of the 2024 season. Although, with Sinner *not* competing at the Olympics that placed Alcaraz as the *dual* Grand Slam Champion of the season to go on to claim Gold – coming incredibly close. Despite **Djokovic**'s record of the season, he lifted his Level when it mattered and so Gold and Silver went the way of the rankings without surprise other than the sheer **Level of play** between

Djokovic, ranked No. 2 behind Sinner at the time, and Alcaraz ranked No. 3. Similar to the ATP Tour, the WTA Tour was absent of one of its primary *two* at the Olympics with Sabalenka not playing making Swiatek the overwhelming favourite to *win* Gold, however, Gauff was ranked No. 2 at the time and presented the greatest hurdle as did Paolini who made *dual* Grand Slam *finals* leading into Paris 2024. But that wasn't to be. The Gold anomaly was Zheng – the AO *finalist* who had not been tracking since her peak earlier in the season, but went on to *win* against Swiatek in the semifinal before facing Vekic for Gold. Interestingly, Vekic – ranked 21 after her 2024 Wimbledon peak, faced Schmiedlova – ranked 67, in their semifinal whilst the other semifinal was between *two* Top 10 players. There is a similarity here across both Tours as Musetti – ranked 16, played Auger-Aliassime – ranked 19, for Bronze and it was Musetti *outside* the Top 10 to claim the third *medal*. As for the women, Zheng was *not* expected to *win* against Swiatek but she was the sole Top 10 player in the *final* – favoured to *win* over Vekic, who was the player outside the Top 10 to medal alike Musetti, but claimed Silver ahead of Swiatek's bronze. Noteworthy, alike Azarenka at the 2012 Olympics, Swiatek also claimed Bronze as No. 1 in the world.

No matter how you look at the Olympic results historically, there are *key* patterns and then there are sheer anomalies. But more than anything, a Top 10 ranking only holds *less* than half of its weight at an Olympics over the course of the last decade. The 2012 Olympics was when these rankings *and* a player's Grand Slam success held significant weight, but not since then has there been the same Level of consistency in *play*. The *two* active *best* players in the world on the men's side *did* fight for Gold, whilst the *best* women's player in the world had to be overcome by the eventual Gold medallist. But if there's one thing for sure, if you're in a performance lull earlier in the season and then *your* peak is suddenly switched on, and it also happens to be an Olympic year, then by all accounts ride the *high* because alike **Djokovic** *and* Zheng, their peak performances had not been consistent throughout the season until mere weeks before when both of these players in their own right found the Level of play that was responsible for *winning* Gold.

When considering the Final 8 of Paris 2024 both the women's and men's Tours achieved a 50% metric of players ranked inside the Top 10 to advance to the quarterfinals. On the WTA Tour, Swiatek, Zheng, Krejcikova and Collins held Top 10 rankings – 50% to progress through to the semifinals, however, both Swiatek and Zheng were on the same side of the draw. This meant the remaining

50% that were not ranked inside the Top 10 – Vekic, Schmiedlova, Kostyuk and Kerber, had surpassed *all* seeded players on their side of the draw with Kerber being the sole Grand Slam *Champion* and former Olympic medallist who pushed Zheng to *three* in the quarterfinals and *almost* came within reach of the *win*. On the ATP Tour, Alcaraz, **Djokovic**, Zverev and Ruud held Top 10 rankings with an identical metric – 50% to advance past the quarterfinals. Tsitsipas, Paul, Musetti and Auger-Aliassime were those *not* ranked inside the Top 10 but all *four* of these players were ranked inside the Top 20 with Tsitsipas the sole former Grand Slam *finalist*. Unlike the women's draw, both Top 10 men's players – Alcaraz and **Djokovic**, were on different sides of the draw that allowed the *two* remaining Top 10 players to contend for the Gold. Of all the Final 8 players across both sides of the Tour, *only* Krejcikova and Zverev passed the threshold *twice* leading into the Olympics, whilst Alcaraz and **Djokovic** passed the threshold *three* times – at *all* Grand Slams prior to the Olympics. Swiatek, Zheng, Vekic, Kostyuk, Musetti, Tsitsipas, Ruud and Paul all crossed the threshold *once* prior to the Olympics – 50% per Tour, whilst Collins – despite her Top 10 ranking at the time, Schmiedlova, Kerber and Auger-Aliassime had *not* progressed to the Final 8 prior to the Olympics this season.

Four of the *best* players in the world medalled at the 2024 Olympics and yet the Final 8 threshold had mixed results. If anything, 50% of the Final 8 across both Tours held a Top 10 ranking – *not* the highest metric of the season but also *not* a regression. *Both* the No. 1 seeds medalled, *two* Grand Slam winners this season medalled, *two* Grand Slam *finalists* this season medalled plus *two* players who were riding peak performances medalled – a progressive metric that whilst it is not equal to that of London 2012 it comes close. The results of the 2024 Olympics adds weight to the cumulative Final 8 across the season and what is most telling is who showed up, who opted *not* to play and who will *maintain* their Level of play for the rest of the season. Of these players, the question begs *how* many thresholds were crossed over the course of the season and if the set benchmark of *two* favours those who have gone on to *win* Grand Slams and/or Olympic medals or is this threshold a *key* metric to simply further track the 8% – this season and those to edge closer in the ones to come.

2024 US Open

Six of the seven best players on the WTA Tour progressed to the Round of 16 with Rybakina being the only player inside this ranking range *not* to progress. The remaining *three* Top 10 players – Krejcikova, Sakkari and Ostapenko all fell in earlier rounds. *Four* of these six players progressed to the

Final 8, however, the two players who fell – Gauff and Paolini, lost to Navarro and Muchova, respectively. What is important to note here is that Navarro has been favoured to become a barrier breaker for most of the season and this result was simply another *peak* performance from Navarro who had been following a progressive trajectory – another telltale sign of a potential barrier breaker. As for Muchova, she in fact progressed to the semifinals *last* season and was only without a seeding due to *not* playing for the first half of the year when in fact her Level of play almost explicitly aligns with the 8 Keys and has reaffirmed that it is only a matter of time until Muchova becomes a barrier breaker once again.

The players of the Final 8 on the WTA Tour were partially anticipated. *Still* the No.1 player in the world at the time, Swiatek had a significant season with limited rest periods in between and for her to progress to the quarterfinals was a positive data point to accompany her FO *win*. Swiatek was to play Pegula in the quarterfinals who has been susceptible to regress outside the Top 10 for most of the season – *not* playing the FO and making the 2nd Round at both the AO and Wimbledon, this was Pegula's best Grand Slam result since *last* season. What is important to consider between Swiatek and Pegula's quarterfinal is that Pegula had *more* matches on hard court under her belt than Swiatek leading into the 2024 US Open which was only a little more than *three* weeks post Olympics and Pegula had *more* rest periods throughout the season over Swiatek. Considering these external factors, Pegula's straight sets *win* over Swiatek was undoubtedly *almighty* yet equal parts a telltale sign of a player who was suffering from over-training and/or overloading. Pegula had the *fresh* legs and the momentum on her side, this was her greatest match of the season, highlighting why Pegula *should* remain inside the Top 10 *if* she can maintain this Level of play.

When Zheng and Sabalenka played their quarterfinal match – a repeat of this year's AO *final*, the primary concern was whether Zheng had enough time to recover, alike Swiatek, but from her Olympic triumph. Noteworthy, since Zheng's AO *finalist* result, her season was *less* than impressive with a 3rd Round at the FO followed by a 1st Round exit at Wimbledon. These results did place Zheng on the backfoot and yet her Olympic Gold appeared to bring back the form that allowed Zheng to progress to her maiden Grand Slam *final*. In contrast, Sabalenka was well rested having opted *not* to play the Olympics and was also absent from Wimbledon – a significant rest period for any player during the season. Although, with the AO title to her name and a quarterfinal result at the FO, Sabalenka *was* favoured to progress to the *final* after her successful lead up events that included *wins* over both Swiatek and Pegula. Factoring rest again into the equation, Sabalenka's straight sets *win* over Zheng was not surprising, although unlike Swiatek, Zheng had *not* been in form for most of the season, while Sabalenka was on yet another *peak*, starting to edge closer towards Swiatek's hold at the top of the game.

Next came Navarro and Badosa – who had started the season outside the Top 100 after a steep rankings regression and had managed to claw her way back to be *seeded* at the US Open. Along with Navarro, Badosa was *one* of only a handful of players this season to be on the radar and tracking towards the Top 10. But Navarro was on a firmer trajectory towards the Top 10 and after her *win* over Gauff, this underscored Navarro's ruthless form. With a 3rd Round result at the AO, followed by a Round of 16 at Roland Garros and then a quarterfinal at Wimbledon, Navarro continued her progressive performance by making the semifinals of the US Open *with* the *win* over Badosa. The remaining quarterfinal was between Muchova and Haddad Maia – despite being seeded, was the *surprise* player of the Final 8 due to *not* tracking as noticeably as any other Final 8 player. Of interest, Haddad Maia was ranked inside the Top 10 briefly during the 2023 season but has primarily remained *around* the Top 20 ranking range this season. Nonetheless, Muchova was the stronger player, having only recently regressed outside the Top 10, to advance to *another* US Open semifinal alike last season.

The semifinals had the *two* top remaining seeds – Sabalenka and Pegula in their respective halves with *both* progressing through to the *finals* but neither Navarro nor Muchova went down without a fight. Navarro pushed to a tiebreak in the second against Sabalenka whilst Muchova ran away with the first against Pegula, but ultimately *two* of the in-form players leading up to the US Open were the *two* left standing. As for the *final*, the result was as anticipated per ranking, but Pegula's Level of play was *almost* within reach as she showed signs of why she was able to get the *win* against Swiatek and come close to almost taking a set from Sabalenka. Nonetheless, this would be Sabalenka's *second* Grand Slam of the season and the *win* by far put Sabalenka within reach of ending the year as the No. 1 player should she *maintain* this peak and continue to hold it for what's left of the season.

On the ATP Tour the numbers were incredibly similar with *seven* of the Top 10 players progressing to the Round of 16 and of those players *five* would make it to the Final 8. But unlike the WTA Tour, only *one* of these players progressed to the semifinals and eventually, *finals*. Of those to make the Round of 16, Sinner secured his place with an impressive *win* over Paul who remained on track to become a barrier breaker early *next* season, then it was Medvedev with a ruthless *win* over Borges, although this was Borges' *second* Round of 16 result this season – enough to progress his ranking inside the Top 30 and potentially place Borges on track to join the Top 20 *next* season. Draper followed, after securing a *win* over Machac, and his appearance *was* the anomaly of the quarterfinals, although Draper's *key* patterns of play signalled his arrival. Then came De Minaur with his *win* over Thompson and De Minaur continued to show *why* he should be firmly ranked insight the Top 10. Next was Fritz over Ruud, followed by Zverev advancing past Nakashima. The final Round of 16 matches were between Rublev and Dimitrov who progressed, and Tiafoe and Popyrin – another surprise result, but after playing some of the best tennis of his career shortly prior to the US Open, this was a healthy metric for Popyrin to illustrate a *new* baseline had been formed despite Tiafoe being the player to advance.

Considering the Final 8 on the ATP Tour, *five* of the *eight* held a Top 10 ranking and only Sinner advanced with perhaps the toughest quarterfinal of all, against nonother than Medvedev, whilst De Minaur had perhaps his best chance to make his first Grand Slam semifinal but it was Draper who was simply the better player with De Minaur carrying an injury. Fritz played some of his *best* tennis against Zverev for a repeat of his result at Wimbledon with this *win really* being a turning point – a progressive metric for Fritz having already made *two* quarterfinals this season. Tiafoe notched a *win* over Dimitrov which was yet another surprise with the *lower* ranked player progressing in *three* quarterfinals whilst Sinner was the only *higher* ranked seed to progress.

Of the *three* players in the semifinals with the lower ranking, Draper firmly placed himself on the radar as a future barrier breaker if he can continuously deliver *this* Level of play and build upon it *next* season, irrespective of his loss to Sinner. As for Tiafoe and Fritz, both players have previously ranked inside the Top 10 but Fritz is the player with greater consistency in his results *and* in maintaining a ranking either inside the Top 10 or slightly outside. Whilst their semifinal encounter was pushed to *five*, if anything this highlighted Tiafoe's capacity to progress once again closer towards the Top 10 *if* he can sustain this *Level* of play throughout the season in contrast to periodic peaks. This is a pattern that Fritz has previously succumbed to with intermittent peak performances in contrast to establishing a peak baseline and *then* building upon it throughout the season. This season, however, is the *first* time Fritz has managed to maintain his *peak* baseline and it has been his most consistent *peak* sustained throughout his career. As for the *final*, it was Fritz's maiden Grand Slam *final* whilst Sinner was simply too good, going on to claim his *second* Grand Slam of the season to follow the achievements of Sabalenka with both 2024 AO Champions claiming the 2024 US Open title.

Insofar as player rankings after the US Open are concerned, there were notable changes across *both* Tours although only *one* barrier breaker per Tour – Navarro on the WTA Tour and Fritz on the ATP Tour. Interestingly, per the predictive analytics that had been tracking Navarro this season, her arrival inside the Top 10 was *not* a surprise. With Fritz's *finalist* result, this allowed him to reclaim his Top 10 ranking which he held at the beginning of the season *and* with his consistent *high*-Level results at Grand Slams throughout the season, this was anticipated. Fritz's Top 10 ranking did, however, come at the expense of De Minaur which does not track granted his *three* consecutive quarterfinal results at the Grand Slam Level. In respect to Navarro, her new Top 10 ranking came at the expense of Ostapenko who had been in and out of the Top 10 throughout the season. Other movers inside this ranking range included Pegula progressing to *three* which does not align with the results this season when Paolini progressed to *two* Grand Slam *finals* and *two* Round of 16's only to be surpassed by nonother than Sabalenka. Gauff regressed slightly due to *not* defending her title whilst both Krejcikova and Sakkari were pushed back slightly. Similarly, Hurkacz, Ruud and Dimitrov were pushed back by Fritz although Rublev remained at six which does *not* correlate with Grand Slam results this season, whilst **Djokovic** slipped to *four* and Zverev progressed to *two*.

If anything, the 2024 US Open did not necessarily correct the rankings as these had slightly aligned *prior* but the cumulative results of the Final 8 highlight the *continued* discrepancies across both Tours. The rankings also highlight *after* all Grand Slams had been played, both Tours were closer in alignment in contrast to the beginning of the season. Interestingly, the biggest movers are those *outside* the Top 10 but *inside* the Top 30 with significant traction made throughout the season for several players across both Tours. And for each player progression and/or regression, there is a story that has unfolded – *how* a player's performance across the season can be *predicted* from their cumulative Grand Slam results *and* their Level of integration of all 8 Keys. If one or more of these *Keys* goes amiss, the results will show. If one or more of these *Keys* becomes more solidified in a player's game, the results will show. And the best part becomes clear when players like Swiatek, Sabalenka, Sinner, Alcaraz and Krejcikova *further* integrate these Keys and add to their respective Grand Slam *wins* with *all* Grand Slam *winners* across the 2024 season *already* Grand Slam Champions in their own right, or in Sinner's case, *after* the 2024 US Open.

Final 8 Deductions

At the beginning of the season Swiatek was leading Sabalenka by *less* than one thousand points then post AO this gap ironically *increased* despite Sabalenka claiming the Championship. By the end of March, this lead had increased to almost *two thousand* but was aligned with Swiatek's arrival for the season – a slower start than Sabalenka but one that would see Swiatek dominate for the next *three* to four months. As a result, Swiatek extended her lead by more than *three thousand* points by the end of June and come the end of July this gap was slightly short of three thousand with Sabalenka gaining ever so slightly. At the end of September, this gap was down to less than *two thousand* and by the end of October, due to Swiatek *not* playing *two* larger tournaments that Sabalenka did play, Sabalenka went on to claim the *number one* ranking due to Swiatek *losing* points and Sabalenka continuing to ride her peak performance.

A similar *rankings* tale can be said about Sinner who was *not* within reach of **Djokovic** at the beginning of the season. **Djokovic** was ranked *number one* and was almost *five thousand* points ahead of Sinner's fourth ranking – prior his *two* Grand Slam *wins*. Even after the AO, despite Sinner's *win* he was still ranked behind **Djokovic** by almost *two thousand* points but Alcaraz was within reach at *two*. Come April, Sinner had 'switched' rankings with Alcaraz and a little more than a *thousand* points separated him from **Djokovic**. Come June, **Djokovic** was pushed to three and Sinner claimed the *number one* ranking despite Alcaraz *winning* the FO. Whilst Alcaraz trailed Sinner by less than a *thousand* points, **Djokovic** lost over a *thousand* points and Alcaraz gained a *thousand* courtesy of his *win* but Sinner also secured over *seven hundred* points – enough to put him in the lead.

At the beginning of August, **Djokovic** was back to *two* whilst Sinner kept his *thousand* point lead. Come October, with Sinner adding *another* Grand Slam to his name, he took more than a *four thousand* point lead – almost equal to the highs of Swiatek mid-season. By this point, Alcaraz reclaimed the number two ranking after Zverev briefly notched the second ranking, before Alcaraz and Zverev switched places again, before **Djokovic** was the player to be pushed back to *four*. Unlike the proximity between Swiatek and Sabalenka for the season ending *number one* ranking, Sinner had now achieved the season ending milestone courtesy of his impressive lead. But the *number two* ranking remained up for grabs with Alcaraz, Zverev, **Djokovic** and even Medvedev within reach of finishing the season *second* in line.

On the WTA Tour a similar *points* margin existed for the *number three* ranking as Swiatek and Sabalenka were trailed by Gauff with less than a *two hundred* point lead over Pegula who has a little more than a *three hundred* point lead over Rybakina. But then there is Paolini only *one hundred* points shy of the Top 5 – by far a misalignment with her results this season, before Zheng trails by slightly less than a *thousand* points – a similar margin that separates Fritz from Medvedev. All these margins, however, are prior to the season ending *Championships*.

On both Tours, the *Championships* proved to be the reaffirming factor to not merely solidify the 8% but highlight the true Top 5. For the WTA Tour, Pegula's withdrawal meant Swiatek lost her chance to claim the *number one* ranking, and for the ATP Tour, it allowed Sinner to add his name to the history books. But for both Tours, the respective Championships aligned with Top 5 with most *recent* seasonal results. Whilst the data suggested that if **Djokovic** could maintain a Level of play similar to that of Wimbledon and the Olympics, he would likely finish as *number three* in the world *if* he remained active and *if* Alcaraz continued his peak form he would finish at *number two* – deservingly so after his *two* Grand Slams this season. But neither was to be the case due to **Djokovic** not playing and Alcaraz experiencing a drop in performance. The fourth and fifth rankings were to go to Zverev then Medvedev with Zverev making *four* of *five* Final 8's this season to Medvedev's *three*. But as **Djokovic** did *not* play and Alcaraz dropped his Level, Zverev edged ahead in the rankings, but as far as Grand Slam results are concerned, Sinner and Alcaraz remain deserving of the top *two* followed by **Djokovic**, Zverev and Medvedev to round out the Top 5. Next comes Fritz with *three* Final 8 results this season with his Level of play correlating with the *sixth* ranking, however, without **Djokovic** in action and *with* a season ending peak, Fritz bumps to the Top 5 and replaces **Djokovic**. Per the data, following these initial *six* players should be De Minaur at *seven* with *three* Final 8 results alike Medvedev. The remaining *three* players of the Top 10 would then be Dimitrov, Ruud and Tsitsipas with *two* Final 8 results each before Rublev with *one* Final 8 result. But that's not how the final rankings of the 2024 season read. It is how the Final 8 distributes each player and their respective performances over the *five* biggest events of the season.

Interestingly, both Paul and Musetti have *two* Final 8 results each – ahead of Rublev, however, Rublev has been able to achieve a relative high Level of play over the course of the season but *not* at the Grand Slam Level. Tsitsipas, Paul and Musetti rank just outside the Top 10 but all *three* of these players are likely to become barrier breakers next season as Rublev becomes more susceptible to regressing and the likes of Dimitrov and Ruud, per Grand Slam metrics, would be next in line to be displaced should they lose their grip on the 8 Keys and subsequently regress *outside* the 8%.

For perspective, Sinner, Alcaraz *and* **Djokovic** secured *four* of *five* Final 8 results whilst Sinner was the only player of the three *not* to play *one* of the five primary events this season – the Olympics. This reaffirms **Djokovic**'s place inside the Top 5 if not for how the rankings fell in the final weeks of the season. In contrast to these results, Rublev remains the only Top 10 player to appear *once* in the Final 8 that also includes Hurkacz, Auger-Allisime, Tiafoe and Draper – the only player of the *four* outside the Top 10 who has *not* previously been ranked inside the Top 10. These results further support the likelihood of Draper joining the Top 10 *next* season, which pushes the 2% to 3% due to the misalignment of the rankings this season. In turn, these deductions also highlight that only 7% of players will be secure in their Top 10 ranking next season and that the order of alignment differs between the Final 8's accomplishments *and* the season ending rankings – a consistent pattern throughout the season.

By equal deduction, Swiatek and Sabalenka had *three* Final 8 results this season, however, Swiatek played all five events to Sabalenka's three – skipping both Wimbledon and the Olympics. This highlights why Sabalenka has been able to finish the year on a peak and why Swiatek's performance started to drop. The only other players of equal standing are Krejcikova and Zheng with *three* Final 8 results each and Krejcikova the only other players to *win* a Grand Slam this season, and Zheng the player to *win* Olympic Gold. These are the primary four players of the Final 8 this season, however, both Gauff and Paolini progressed past the Round of 16 at *all* Grand Slams this season which is unmatched by any other player plus *two* Final 8 results. Due to this Level of maintenance, Gauff and Paolini should be ranked *three* and *four*, respectively, before Zheng – who did not consistently *maintain* her peak performance until the last quarter of the season to round out the Top 5.

Due to Krejcikova's Grand Slam success *and* Final 8 results this season, however, a ranking behind these initial *five* would align with the collective results irrespective of the season ending rankings.

These initial *six* players are closely followed by Rybakina with *two* Final 8 results, however, due to these results being in the first half of the season, this slightly pushes Rybakina back due to these *peak* results not being as equally dispersed across the season. In contrast, Paolini's results were dispersed across the season and Zheng managed to peak at the beginning *and* throughout the final *quarter*. Whilst Gauff 'showed up' throughout the season and claimed the season ending *Championship*, Paolini made *two* Grand Slam finals – ahead of Zheng and Pegula who were the other *finalists* of the 2024 season. Following Swiatek, Sabalenka, Gauff and Paolini, the likes of Zheng, Krejcikova and Rybakina follow at a slightly different Level due to their respective successes over the course of the season. What is important to note is that Krejcikova is ahead of Zheng insofar at the Final 8 are concerned and Grand Slam metrics, but due to Zheng's strong season finish, this does place her ahead on *non*-Grand Slam performances.

Next in line 'should' be Navarro, per the data, as not only did Navarro become a barrier breaker this season, she has *two* Final 8 results to her name. Noteworthy, Navarro has been tracking *all* season and the predictive analytics continued to pinpoint Navarro's arrival into the Top 10. Then comes Pegula courtesy of her hard court *swing* and *finalist* peak but remains with *one* Final 8 result. Of importance, Pegula would not align with her Top 10 ranking this season if she did not start to track in the final quarter of the season with *regressive* Grand Slam results in contrast to those ahead of her in the rankings. As for the *last* Top 10 ranking, this is up for grabs – literally. Haddad Maia became a barrier breaker in the later end of the season but remains with *one* Final 8 result, whilst Collins, Muchova, Ostapenko, Jabeur *and* Vondrousova were all ranked inside the Top 10 at some time this season and *all* have *one* Final 8 result to their name. *Two* players who have been tracking this season are Kostyuk and Vekic with *two* Final 8 results whereas Kostyuk is slightly ahead of Vekic in the rankings to put her within reach of becoming a barrier breaker *next* season if she continues her current trajectory. Kalinskaya and Badosa have been tracking this season and are both on the cusp of the Top 10, yet both have *one* Final 8 result to their name. Then there is the sole outlier in Kerber – former *multi*-Grand Slam Champion who turned back the clock at the Olympics for *one* Final 8 result before retiring from the game. The remaining players with *one* Final 8 result include Noskova, Yastremska, M. Andreeva and Svitolina with all players being ranked inside the Top 30 at the time *after* their respective *peak* results, although both M. Andreeva and Svitolina have been the more consistent players of these *four* across the season with M. Andreeva set to potentially become a barrier breaker in the *next* two seasons.

Last are *two* anomalies for the season – Sun being ranked outside the Top 100 for her breakthrough Final 8 result, and Schmiedlova outside the Top 60 at the time of her Final 8 result, but neither have maintained these peaks in the latter half of the season. Whilst the ATP Tour had its anomalies, these players were either ranked inside the Top 20 – Paul and Musetti, had been ranked inside the Top 10 previously – Tiafoe, or had been tracking and held a Top 30 ranking – Draper. What is more telling is that the ATP Tour had only *one* player inside the Top 10 – Rublev, to make the Final 8 *once* with all other Top 10 players at *two* or more *plus* those on the cusp of the Top 10. On the WTA Tour, Pegula is the only player inside the 8% to have one Final 8 result yet Krejcikova, who has *three* Final 8 results was not in the recorded Top 10 rankings *prior* to the WTA Finals but *is* in the data-depicted rankings *and* only after the WTA Championships was Krejcikova ranked inside the Top 10. But that still leaves *one* ranking to be decided.

A consequence of the higher number of players to make the Final 8 on the WTA Tour this season ($n = 26$) in contrast to the ATP Tour ($n = 17$) shows why this last ranking is by no means concrete. If anything, these numbers signal why the identified Top 8 – Swiatek, Sabalenka, Gauff, Paolini, Zheng, Rybakina, Krejcikova and Navarro have such a high Level of maintenance that Pegula needed her *finalist* result to maintain her Top 10 status. But no other player correlates this season due to their high susceptibility. As such, whoever is ranked *ten*, per the data opposed to the season ending rankings, will remain susceptible to regressing outside the Top 10. In the final weeks of the season, Collins, Haddad Maia *plus* the new arrival of Kasatkina – *without* a Final 8 result this season, have all traded places for the *final* Top 10 ranking. Noteworthy, all three of these players have previously held a Top 10 ranking whilst Kasatkina, who did *not* secure a Final 8 result this season, was the player *post* WTA Championships to finish inside the Top 10 – a clear outlier.

The difference between the Tours is apparent in their Level of consistency and the dominance of the Top 10. The ATP Tours Top 10 dominated the Final 8 75% or *more* of the time, except for the Olympics at 63%, during the 2024 season including Hurkacz and Tsitsipas who were ranked inside the Top 10 at the beginning of the season. In contrast, the WTA Tours Top 10 comprised of 50% or more of the Final 8 with the US Open coming in at 88% including players to have held a Top 10 ranking *this* season. As such, the Final 8 is in fact a fair representation of the 50% anomaly on the WTA Tour this season of the Top 10 and *how* a partial alignment formed, however, as shared in the data-depicted rankings order, this was not identical to the *actual* recorded rankings. These corrections took place throughout the season with players ranked inside the Top 10 at the beginning of the season being bounced out whilst new players arrived – to also be bounced out, before the likes of Navarro would confirm the Top 8.

The ATP Tour did have a 50% anomaly in *performance* earlier in the season, however, this was soon corrected and narrowed to highlight the true 2% and respective rate of susceptibility across the 10%. This refined margin was courtesy of more players with the 8 Keys dispersed just *inside* the Top 10 and just *outside* the Top 10 vying for a higher Level of *maintenance* to solidify their place inside the 8%. Interestingly, the WTA Tour had the opposing problem – lower levels of *maintenance* of the 8 Keys that continued to push players *inside* the Top 10 *outside* the Top 10 and why the *final* ranking remains unclear. This was made even more problematic with the *eleven* additional players to make the Final 8 this season not ranked inside the Top 10 plus the *eight* additional players who have held a Top 10 ranking this season *excluding* Paolini, Zheng and Navarro who are first time barrier breakers this season *and* are included in the 8% courtesy of their results. Noteworthy, of these *eight* players – Krejcikova, Collins, Muchova, Ostapenko, Jabeur, Vondrousova, Haddad Maia and Kasatkina – *all* have previously held a Top 10 ranking prior to the 2024 season. *If* one player had to be selected to join the Top 10, irrespective of the recorded rankings, the data-points towards Collins due to her results this season, although as Collins experienced a slight regression in the last quarter of the season and wasn't as active, Kasatkina managed to collect enough points to *push* ahead and claim a Top 10 ranking – for the *first* time this season complemented by a late season *peak*.

Eight by Three

Of the *forty* Final 8 places across each Tour, the ATP Tours Top 10 dominated with *seventeen* players collecting all places in contrast to the WTA Tours *twenty-seven*. Across each Grand Slam and Paris 2024 there was a story to be told from those anticipated to even **Djokovic** not having won a title all season and then going on to *win* the elusive Gold. Sabalenka played a 'smart' year with her rest periods paying dividends for the later end of the season as she secured the *number one* ranking. Swiatek suffered burnout and needed rest which allowed Sabalenka to keep firing whilst on the ATP Tour there was a battle for *number two* as Sinner surged out of reach. On the WTA Tour, Gauff, Pegula, Paolini, Rybakina and Zheng had the chance to secure *number three*, but Gauff's *Championship* triumph amplified why she really was close behind both Swiatek and Sabalenka *all* season despite not progressing to a Grand Slam *final* or two, alike Paolini, who remains close behind.

It was a *huge* season. It remains daunting and unrelenting. If anything, the Final 8 showcased what the best in the world *need* to do to remain inside the 8% whilst highlighting the discrepancies across both Tours. It became clear that the ATP Tour was in relative alignment, aside from a few exceptions *and* new barrier breakers soon to make their mark next season as the 2% becomes more susceptible to their arrival. In contrast, the WTA Tour showed greater levels of inconsistencies. Whilst the data-driven rankings shared *how* the rankings 'should' appear, that was not the case for the recorded rankings. And this identified a known variable of player performances *and* rankings that needs to be considered when vying for the top of the game – *your* Top 10 ranking or Grand Slam *success*. What once worked in unison is no longer guaranteed. But the real *secret* is in delivering on both sides and over time you will establish a profound Level of maintenance that will place you at the helm of the game alike **Federer**, **Nadal** *and* **Serena** did throughout their careers, and as **Djokovic** continues to do so. Sabalenka, Swiatek, Sinner and Alcaraz lead the *new* charge as the primary four across both Tours, however, time will tell whether their respective levels of maintenance will be upheld over the next decade. But as each of these *four* remain as the frontrunners of their respective Tours, courtesy of the integration of the 8 Keys, Gauff closely follows whilst **Djokovic** remains in contention to make *more* history – *both* Tours have fierce competition.

Whilst Sabalenka and Swiatek were in the spotlight all season followed by Paolini with her dual *finalist* results, it was Gauff who was *almost* as consistent and why it was not surprising that Gauff, ahead of both Sabalenka and Swiatek, finished the season as the WTA Finals *Champion* – defeating the two along the way. Although Paolini couldn't perform at the Level she had consistently produced across the season, Paolini made the Final 8 of both *singles* and *doubles* – a remarkable feat and a substantial year. Despite no Grand Slam or *finalist* result for Gauff this season, her season ending *Championship* result highlighted why Gauff remains on track for *another* Grand Slam and the number *one* ranking in the seasons to come. Behind Gauff at *four* is Paolini courtesy of her standout season. Zheng closely follows Paolini at *five* due to a breakthrough final *quarter* that included advancing to the WTA Finals. Insofar as the data is concerned, Paolini remains a *strong* four whilst Zheng's late season surge places her in serious contention for a Grand Slam next season. These *five* players lead the pack over the 2024 season closely followed by Krejcikova who surpasses Zheng in Final 8 results *and* consistency of results across the duration of the season. With such a close margin between these players, next season is set to impress as whilst *two* of the Top 5 claimed Grand Slams this season – 50% if pushed to the leading *six*, all are Grand Slam Champions and/or *finalists* with Rybakina and Pegula in the mix.

Zverev takes out *third* in line on the ATP Tour due to both Sinner and Alcaraz claiming *two* Grand Slams a piece this season despite the recorded rankings placing Zverev at *two*. With Zverev joined by Medvedev, **Djokovic** and Fritz as Grand Slam *finalists* this season, his Level of consistency across the season places him at the forefront, however, both Fritz and Medvedev still had strong all-round seasons. Of course, **Djokovic** did in fact have an impressive season, but it was not at his typical Level nor at the Level of Sinner or Alcaraz and due to a reduced schedule, Zverev, Fritz and Medvedev benefited in the rankings *and* by upholding their respective Level of maintenance of the 8 Keys. Fritz is bumped ahead of Medvedev courtesy of his ATP Finals result where he edged out Zverev to progress to the *finals* whilst Medvedev remains in the Top 5. Whilst **Djokovic** has been pushed out of the Top 5, this is not directly correlated with his performance due to the lack thereof in contrast, but Ruud is the player between – the Top 5 and **Djokovic** due to his successful clay season *and* strong finish to the season.

With a significant misalignment between the rankings over the course of the season on both Tours, the *final five* per Tour – closely followed by *six* then *seven*, depict *how* the 2024 season played out. As for the remaining *five* or *three*, these are relatively true, however the margin of the 2% remains under threat. Whilst the 8% has finally aligned with the rankings and it took *all* season for the correction to take place, the 2% does in fact remain but is more vulnerable courtesy of *more* players on the brink of becoming barrier breakers next season. This places heightened pressure on those who have not locked in **eight by three** and the frequency remains *less* in those outside the leading *six* and *seven*, respectively. Unequivocally, Sinner is the sole player of the primary *two* per Tour that stamped ahead this season, as unlike Alcaraz and Sabalenka, he added the season ending *Championship* to his tally, and alike Swiatek *last* season, will start the 2025 season with a significant lead. But do not let this be mistaken for *not* within reach. Alike Swiatek's lead at the beginning of the 2024 season, it was Sabalenka who finished on top courtesy of her *maintained* peak in the latter half of the year and if Sinner is to be overtaken, **eight by three** will place that player in contention for Grand Slam titles *and* to battle it out for a place in the primary *two* or be it *three*.

Despite the season ending *Championships* not incorporated into the Final 8, the respective results predominantly reflect the rankings corrections. From the *finalists* – Zheng and Fritz, to the *Champions* – Gauff and Sinner, these tell a tale of peak performances sustained and confirmed the Top 5. The role of **eight by three**, however, has never been as clear from *how* the Final 8 matters, *why* the 8 Keys are a must and *what* the 8% really means for a player's rate of success and to finish the season with a *bang*. Those who harnessed **eight by three** and remained tall and/or missed out due to mere margins, are the ones who pose the greatest danger – from the primary *two* per Tour and those that closely follow by whispers. The players that are yet to embrace **eight by three** *will* become more readily displaced next season as the gap widens and room becomes available for *new* players to arrive, to challenge the 8% and demand *more*. Those relentless enough to keep **eight by three** at the forefront and put in the work will maintain their hold inside the 8% but those who miss the mark of knowing the power of **eight by three** will lose their hold inside the exclusive ranks.

In contrast to the Round of 16 benchmark set throughout the *Collection*, **My [Tennis] Black Book** is the first to deduce the Final 8 and deliver new and additional performance insights. The Final 8 was built for *you* to understand what it takes to not simply *navigate* towards a Top 10 ranking, but *how* the collective results across the Final 8 over the course of a season can also depict who will also ascend into the Top 10 this season *or* be it next. Similarly, the Final 8 charted those despite being ranked inside the Top 10, are the ones to take advantage of during play – these are the more susceptible players who have a lower Level of maintenance of the 8 Keys and these *wins* will help put you on the map – if you can *plan* accordingly. Although it is important not to mistake the Final 8 findings with the *impact* of Round of 16 metrics – something Gauff, Paolini, Sinner, Zverev, Fritz and Medvedev achieved in contrast to Sabalenka, Swiatek, Zheng or Alcaraz when considering the Top 5 per Tour. For more on the Round of 16 benchmark, that's where *stages* two and three come into play.

But here and now, the lessons of the Final 8 provide a new metric and powerful at that, to consider what's factually possible when all *three* **eight** 'markers' are in harmony.

Irrespective of *your* ranking and/or *your* Level there is a Final 8 at each event *you* play and *how* you build your baseline and continue to track matters. Players who peak at the beginning *then* end of the season do not have a reliable baseline. Players who peak in one Block per season do *not* have a reliable baseline. Players who progressively build on their baseline over the course of the season – from one baseline to their next peak performance, are the ones who understand the integration of the 8 Keys, their continued maintenance and solidification into their game, and these are the players who *will* track, no matter the Level, and steadily progress towards the Top 10 rankings. The question that remains is whether you choose to be that player – the one who steadily tracks, or the one who opts to periodically peak. Each has its own benefits. But only one has a Level of longevity inside the Top 10 and a Level of established maintenance that can reap the enduring rewards of continuously advancing to the Final 8. It should come as no surprise then, that these rewards are directly correlated with the 8 Keys – happenstance if you will, but the Final 8 courtesy of the 8 Keys is your ultimate reward. Together, by learning to build that baseline and *plan* for that *next* peak performance, the Final 8 will become a reality. But the Final 8 is cyclical with the 8 Keys – the further each of these Keys reaches a Level of autonomy and maintenance, which later will increase in frequency, will remain as a *key* divider between Final 8 results, Grand Slam *finalist* results and ultimately *your* Grand Slam *win(s)*. As for your ranking, the greater *your* baseline and *your* Level of maintenance of *all* 8 Keys, the greater the rate of ascension towards a Top 10 ranking and to not only enter but take hold of the 8%. Now that's **eight by three** in action – the untold story *until* now.

Your Season Wrap

What a whirlwind *or* perhaps rollercoaster. *Your* Black Book has finally made it to the end of *your* season traversing 12 *full* Blocks – *forty-eight* weeks in total and if you've ensured the additional *four* weeks have been dispersed throughout, then that's an entire calendar year *accomplished*. That's **HUGE**. A *congratulations* seems insufficient for *all* the work you have done *and* achieved over the course of the season. If you've progressed to this point with *your* Black Book bristling at the seams, I am so incredibly *proud* of you!

No matter if this is *your* first season on Tour as a professional or *your* first full season enrolled in *your* local Club's high performance program or somewhere in between – completing *your* Black Book over the duration of the season is significant. Not every player will have showed or developed this Level of dedication, discipline and determination – the 3D's you have come to learn and *how* these drivers afford you that extra advantage. But with that, you have now ingrained the habitual process of *planning* towards a Top 10 ranking.

My [Tennis] Black Book has provided you with the tools to *navigate* towards the top of the game as each *stage* provided the foundation for each Level. If you started at Level 1 and *stage* one the journey has just begun – you know what lays ahead. If you started at Level 3 and have covered *all* stages whilst solidifying *all* 8 Keys into your game, then a Top 10 ranking is on the horizon. But recall, it won't happen overnight. The relationship between *your* Level of development *and* performance is synonymous with your rate of progress. Put simply, if you're based at your local Academy and fast-tracked your way through the *Collection* without integrating the *key* learnings and respective components of each Book, then the Top 10 is *not* the warranted outcome. Your time is not yet. The Pathway exists for a reason and whether you've put in *your* initial decade of play, or more, whilst harnessing *all* 8 Keys, the *gradual* load is pivotal to *your* end point.

But let this be a reminder that *this* is your first season of many to come. For those of you who have started on Level 1 and *stage* one let this mark a fundamental milestone in *your* playing and performance trajectory – the roadmap has been paved for *you*. This means *stage* two beckons and then *stage* three before Level 2 is to be commenced. Time is on your side. For those of you who have now become familiar with *all* 8 Keys and the entirety of the *Collection* whilst putting in the work to reach a new Level of autonomy – *your* ascension awaits along with, not *if* but *when*, **eight by three** is your own progressive seasonal milestone and *marker*.

Information Overload, Condensed

Irrespective of *your* Level and *stage*, **My [Tennis] Black Book** encapsulates *key* **Topics** that you have had the chance to consider and embed in your personal 'playing' pedagogy in addition to your weekly **Chapter** segments. The **Topics** provided every *first* and *third* week of *your* Black Book afforded a condensed review of one of the many detailed **Chapters** across the span of the *Collection*. Primarily pulled from *'The Secrets to Optimal Performance Success'* – *stage* one (58%), followed by a selection from *'The Secrets to Optimal Coaching Success'* – *stage* three (29%), *before* providing timely *analyses* of Top 10 *rankings* and the 2024 season (13%), the variation of **Topics** were designed to solidify your foundational baseline whilst providing initial scope for those ready for *more*. Oftentimes, the respective **Topics** overlapped with the subsequent weekly **Episodes** to ensure your initial baseline was firmly intact and where there was no overlap this served as a *new* singular **Topic** introduced to **My [Tennis] Black Book** to add to your knowledge bank.

All *twenty-four* **Topics**, including A Final Lesson, provided *two* additional readings per **Block** to counter the likelihood of *information overload* whilst still providing ample learnings to *navigate* over the course of one **Block** to the next. Kicking off with Hiring and Firing Your Coach plus a brief overview of Rate of Progress and Balance, all *three* of these Topics play their own role in *your* development as a player – long term. Functional Movement was then touched on with a review of *your* C.O.G and R.O.M before Power and Balance followed. Each one of these technical considerations are instrumental in the integration, application and fine-tuning of The V by Dr B © – embedded in the 8 Keys.

Next was Single Sport Advantages that looked at the role of team sports and *how* they can benefit *your* performance, including engagement, before Relearning Discrete Skills touched on injury prevention and the concerns around *rapid-fire* results – fast-tracked progressions. The Role of Education provided a different take on *learning* and current frameworks (*if* applicable), followed by a look at Your Purpose and ensuring you're training with intention. *How* to Learn from Mistakes discussed the concept of 'rewiring' to building behaviours, and then Hydration featured various 'replenishments' to fuel performance whilst briefly dispelling related myths. Knowing when to Take Time Off and the balancing act was then discussed whilst Injuries and how they Correlate with Technique provided an important reminder for load management and the importance of aligning *your* technique with the properties established through The V by Dr B © and its direct correlation *with* a Top 10 ranking and *more*.

The role of Planning was then moved onto and highlighted the premise of *your* Black Book and *how* to continue these lessons for the future whilst Happiness away from tennis reflected on why balance on *and* off court remains *key*. Fair Play and Ethics then looked at the importance of integrity irrespective of the result before *another* take on discrete skills in Fundamental Skills emphasised the direct alignment between *functional* skills and *how* they help to mitigate injuries. These **Topics** cleared the way to introduce Problem Solving and the conditioning of behaviours to the role of anticipation, which was followed by Reading and Performance that provided yet another overlap with the design of *your* Black Book.

For a change of pace, the Power of Song encouraged music to your ears – literally, whilst Age Relevance touched on its *irrelevance* when performance *wins*. This was followed by Persistence v. Walking Away that stressed the 'loop' and your team. Gender Interactions were subsequently discussed and highlighted the advantages of 'mixed' play before a *mixed* **Topic** was shared in Systems, Promises, Power and Luck – condensed under the one *week* to mark the *end* to the *secrets* and their respective intersections. Next were the deductions of results *and* rankings across the 2024 season that included How Rankings are Not the Whole Truth and a multifaceted discussion – From Clay, the Primary Two and the Power of Rest, before A Final Lesson provided the *final* **Topic** to be shared over the course of your *final three* weeks.

Collectively, all *twenty-four* **Topics** acted as a stepping stone – a progression, in their own way whilst initially walking you through the earliest *stage* of the *Collection* before entering the later *stage* of Level 1. The culmination of **Topics** shared ended with *real* and *direct* player insights from points in time during the 2024 season offering a window *after* Level 3 – the 'State of Play' of the best players of the season. These insights shared *how* you can now use the dense data in *your* Black Book to join the field. But that's not all. To deliver on a *new* finding, exclusively in *your* Black Book – adding further weight to its value *and* to add to your now overflowing knowledge bank, the **Final 8** was the *finale* to be shared. Through the **Final 8** an assortment of *new* findings was laid bare exclusive to the 2024 season, providing another Level of scientific rigour to substantiate the interwoven value and need for the 8 Keys if a Top 10 player is to *maintain* their ranking inside the 8%. It was also made clear through these findings that whilst a Round of 16 result *twice* or more per season is a strong and important data metric, making the **Final 8** *twice* or more added *new* data to player peak performances and their ability to either *maintain* these peaks and/or build on them throughout the season.

Whilst the **Final 8** delivered A Final Lesson, it remains an extension of the *Collection* as a whole. The data provided throughout the *Collection* is robust for good reason as each *stage* and Level displayed in Table 2 builds on the one *prior* as the **Final 8** did so in **eight by three**. But as in life, you need to learn to *walk* before you can *run* and this remains true from Book *one* through Book *twelve* – held between your hands, with a *library* of **Topics** to traverse until all 8 Keys are deep-rooted within your game.

By design, **My [Tennis] Black Book** follows that of the *Collection* as a progression made for you. To again counter *information overload*, it has always remained prudent to provide the building blocks of *one key* before the *next* can be shared. After all, it is human nature to *learn* rudimentary concepts before progressing to more complex learnings. On this merit, each Book of the *Collection* follows this pattern, by design, building on its former to provide a *condensed*, albeit detailed lesson – *one* after the other. This format, however, was by no means coincidence. As has been shared throughout the *Collection*, this was always *how* to **Navigate and Plan towards a Top 10 Ranking** was meant to be. Knowingly, this is what the data reiterated courtesy of human nature and *how* each one of us *learns*. By skipping *one* Book or by condensing all *findings* and *learnings* into one tome, would be a surefire guarantee of *information overload* and your ability to retain, let alone apply, *one key* before the next. This would have limited the retention, at best, to a few to reap the rewards of decades of work. By far counterintuitive. But through *intentional* design, knowing *how* to *navigate* and *plan* to truthfully *achieve* a Top 10 ranking has been built for *more* than a few – to ensure each one of *you* has the chance, has the opportunity should you choose, to use each one of these **Topics** – in *your* Black Book and throughout the *Collection*, to lead you towards the top of the game.

Between the Lines

Each week you were presented with *three* or more tasks to keep pace with *your* Black Book from one week to the next. Of those *first* and *third* weeks you had *four* tasks to complete – including the respective **Topic** for that week. For each *second* and *fourth* week, there were *three*. These *three* tasks were *between the lines* each week of each **Block** as they were not as seamlessly before you. Each **Episode** and *named* **Chapter** was to be found, either on the adjoining **Podcast**, 'Beyond Top 10 Tennis' or through the *Collection*. For both there was an element of *navigating* towards each **Episode** and each **Chapter** on a weekly basis before the need to complete your weekly **Plan**. These *three* tasks formed the foundation of *your* Black Book whilst the respective **Topics** afforded additional concept overviews to add *more* to your knowledge bank each fortnight whilst remaining as exclusive additions to the *Collection* – allowing **My [Tennis] Black Book** to take its own place in the extensive lineup of work.

The **Episodes** and **Chapters** made available a *new* means of learning and to understand what it takes to progress towards a Top 10 ranking whilst making available a means to explore *stage* one of each Level. Whilst each **Episode** directly aligns with the initial *secrets*, each **Chapter** corresponds with *stage* one of your chosen Level. This multifaceted approach has allowed you to not only follow the *plan* of *your* Black Book but to *navigate* your way through the initial *stage* of each Level. Whilst **My [Tennis] Black Book** is not designed to span the entirety of the *Collection*, it has been designed to allow you to *track* those initial *stages*. Whereas stage *two* and *three* of each Level is yet to be crossed, that remains up to *you* to follow through until *your* Level has been completed before moving onto the next.

An important consideration for each **Episode** is that irrespective of their alignment with the *secrets*, each Book of the *Collection* is featured in select **Episodes** in no precise order. This format showcases the overlap between Book *one* to Book *eleven* and all that fall *between*. As a result, each **Episode** is somewhere *between the lines* of *one* Book to the next due to the interdependence established across this body of work. **My [Tennis] Black Book** is the first *guide* of its kind to traverse the *Collection* whilst forming its own keepsake – a place to store your own *secrets*, if you choose, for the entirety of the season. As a guide, each **Episode** of *your* Black Book comes with its own message to pen in your weekly *key takeaway* that circles back to your **Plan** and **Chapter** revised for that week. These are the 'Notes', in addition to those of your *fortnightly*

Topic that will provide *key* data-points of your own doing that you can now reflect upon next season and those to come.

To cover the span of the *Collection*, the **Chapters** suggested in *your* Black Book provide a PoI but are by no means comprehensive. What's fundamental to remember is that *your* Black Book has been designed to not merely instil behaviours conducive in developing *optimal* performances *and* results but to act as a bridge between *one* Level and the next. *Between* each Level are various **Chapters** to work through whilst each one listed provides a means to crosscheck that the **Chapters** prior can be marked as *read*. And this is an important reminder – there will be **Chapters** *between* the ones listed at all *stages* and all Levels. If you have followed the designated **Chapters** of each week, you now have a profound baseline to build upon and know what you have *read* and what is yet to make its way into *your* knowledge bank. If you have read *between the lines* in addition to the **Chapters** exclusively named, then you're ready for that next *stage* and/or that next Level. But remember, *your* Black Book is not a race – it is *your* timeline in alignment with your rate of progress.

The culmination of *all* tasks each week was in *your* **Plan** and its subsequent completion. Each of the *twelve* **Blocks** in *your* Black Book provided *four* **Plans** to be completed – *one* each week. Incorporated into each **Plan** was another *twelve* sections for you to complete each of the *forty-eight* weeks – contributing towards your *new* habitual 'State of Play'. The details of these **Plans**, penned by you, are not merely markers of significant milestones throughout the season, but complete the implication *behind* the name – a **Black Book** like no other. This **Black Book** is *yours* and brimming at the seams with your own metrics of the season – what you worked on, what were your biggest motivators to improvements and much more.

A first of its kind, *your* Black Book is a **powerhouse** as is what is penned between the pages – your *superpower* moments included. By working your way through *one* **Plan** to the next, and to now have *forty-eight* weeks of your *own* data is *remarkable*. And I'll be the *first* to tell you that it wasn't easy. To complete each one of your **Plans**, *every* **Episode** and **Chapter** of *each* week *plus* each **Topic** every fortnight was an almighty commitment. The time and effort that has gone into completing *your* Black Book as it was intended is not only the output of an incredible amount of *work*, but wholeheartedly resonates with the 3D's. A surefire signal that *you* are on track.

With the array of data available inside *your* Black Book it now serves *two* purposes, among many, to not simply add to *your* knowledge bank and provide an initial framework to cover *stage* one of Level 1 whilst progressively learning, integrating and applying the 8 Keys to *navigate* and *plan* towards a Top 10 ranking; but your own data has now formed its own **cycle** to *track*. This tracking is in reference to learning from your *own* data – what works for you and what does not, what motivates you consistently that you can use to further enhance your output, *or* how many **Blocks** did it take to build towards that next baseline and that next peak performance. And these are *four* of a multitude of questions you can *now* ask *about* yourself and find the answers right *between the lines* of *your* Black Book. And that right there is *how* courtesy of the multidimensional approach of **My [Tennis] Black Book** – **Topics**, **Episodes**, **Chapters** *and* **Plans**, that *your* Black Book has come full circle – to life, and to mark the end of *your* season.

Work It

Is it time for that extra *push*? That simple *ask* in the **Foreword** has taken on a new light *now* that you have worked your way through the season. Hard work is not for the faint hearted and nor is the pursuit of the completion of a **Black Book** over the course of an entire season. If you've made it this far then you're already ahead of countless others who do *not* have their own **Black Book**, nor have they become familiar with the 8 Keys. This is a **lethal** combination for any tennis player – on *Tour* or aspiring to join the Top 10.

One of the *best* rewards of *your* Black Book is what it has turned out to be – what *you* have created over the course of the season. Each **Black Book** will be unique as it should be – each one of you is an individual and there is no right or wrong. What has always remained fundamental from the beginning is the completion of *your* Black Book and putting in the *work* to see it through. And now comes the best part.

I've been holding back as I penned all *twelve* **Blocks**, and everything laid between to *really* get to talk to you – from *me* to *you*. I wanted to let the work speak for itself and as such removed the *'conversation'* between the embedded *forty-eight* weeks to ensure you really did *work it* before I chimed in. Which means if you've made it this far, I'll repeat what I shared only a few pages earlier – I am so *proud* of you! No matter your ranking – locally based or on the WTA or ATP Tours, there's a significant lesson embodied in **My [Tennis] Black Book** that you are now privy to and have the capacity to *really* bring your dreams to fruition.

Over the course of building the *Collection* not beyond my wildest dreams did I imagine this work would identify *how* to *win* **Grand Slams**, *how* to *win* **more than one Grand Slam** to *'How to Develop a Top 10 Tennis Ranking'* and so much more. If you're yet to complete the *Collection* that's okay – I'll let these endpoints be a surprise. But what this emphasises is that no matter how **wild** your dreams, this work is a testament to *working it* relentlessly.

The Golden Slam, let alone *the* Grand Slam, has seldom been achieved throughout history. The likes of **Federer, Nadal, Serena** and **Djokovic** got close in their own rights, but none of these players (at the time of writing) have been able to accomplish one of these feats. Of course, Steffi Graf is the only player in the Open Era to do so, as shared in the **Foreword**, and fast-forward to *now* and the development of the 8 Keys to sharing **eight by three**, the likelihood of a player to achieve this feat is *increasingly* within reach. But the Golden Slam won't be for another *four* years at the earliest.

My [Tennis] Black Book was timed to coincide with an Olympic year and then it became a *rush* against time. I had to really *work it* to ensure what filled the pages of *your* Black Book was refined and condensed opposed to a much more detailed release than undoubtedly would *not* see the light of day. I knew having been in your shoes that you would be more likely to take *your* Black Book with you – as intended, if it was compact in contrast to an additional weight of its own alike *'The Science of Elite Performance'*. But fitting in *all* that has been encompassed inside *your* Black Book was no easy task whilst adhering to the initial boundary of scale and scope and why the adjoining *Journal* was put together – *your* own **workbook** to more easily complete if you have struggled to fit all your notes within the pages.

With innumerable considerations along the way, *your* Black Book by no means is what it was originally intended, but better – to really **pack a punch**. Of course, in a refined and compact version, as I truly considered each step of the way, whether *I would be prepared to do this if I was in your shoes?* And that's why at the very last minute I realised a *Journal* was a must – in addition to *your* Black Book and all *forty-eight* weekly **Plans**, to include a standalone **workbook** as the much needed adjoining *Journal* to **My [Tennis] Black Book** that you can also take with you *or* leave at home to complete each night before bed, or first thing every morning. If *I* was in *your* shoes and was committed to *work it*, then I'd want that **workbook** – the *Journal* to **My [Tennis] Black Book** that I could scribble all over after I'd jotted down my initial thoughts *here*, to use as an *extension* with simply more room to *flow*.

> I can still recall being in *your* shoes. I was that player who was relentless – asking *how* and wanting to know *more*. I had that insatiable thirst to achieve *the* Golden Slam and I envisioned winning Grand Slams like I would eat my breakfast – *daily*. I would rise before it was time to go to School and *work it*. After School I would continue to *work it*. But my results were never what I had envisioned. No matter how *hard* I worked – no matter how much I *worked it*, there was still a gap that I couldn't overcome. The answers I sought were not available. No matter where I

turned, there was a barrier – whether financial or otherwise. This *feeling* never left me. So, I decided that I wouldn't let this *feeling* remain if there was something I could do to change it – for *you*. And that is *why* I have continued to *work it* – for you, because I know that *feeling* and what it is like to have questions and want to know what's next, *how* to *navigate* towards the top of the game, how to *plan* to progress towards the Top 10 or how to bring *your* dreams to life and *win* a Grand Slam and/or edge closer towards the elusive *Golden* Slam.

It can be done. These answers are now known. If I had to *experience* those barriers then I would do everything I could to help *you* overcome them. With the *Collection* and **My [Tennis] Black Book**, you are better prepared than ever before. But this is the beginning. Whilst *your* Black Book has completed one *full* season, you need to *work it* – from *stage* one of Level 1 through to *stage* three of Level 3 and beyond. This is the beginning as *working it* is akin to that fire that drives you – rain, hail or shine, your core driver remains firmly set on *navigating* towards a Top 10 ranking. And the *planning* enveloped in *your* Black Book from one season to the next is your *guide* to bring your dreams to life.

This is what I yearned for. This is what was *missing* – a simplified *plan* to help guide players towards the pinnacle of play. And now, in conjunction with the *Collection* you have access to what has never previously been available – I made sure of it, as promised, that if I could help, I would. And I've *worked it* ruthlessly so you can overcome the barriers that I faced when I *was* in your shoes. And the *Journal* is in addition to *your* Black Book – ready when you *seek* more space to complete your **Plans**, more room to pen your thoughts, or simply an additional **workbook** to use as an adjoining **diary** when in need.

The data encapsulated in *your* Black Book is yours to prepare for next season and rework *how* you *planned* this season to get even more out of the next. And this is a continuation. Whether you're moving on to *stage* two of Level 1 or are now beginning *stage* one of Level 2, your own data is now available for you to better *plan* and prepare for the next *twelve* **Blocks** ahead. The same applies if you've progressed to the Final 8 this season or further – be that a semifinal result or even a Grand Slam *win*. If the latter, it now comes down to maintenance and achieving *replicated success* – solidify the 8 Keys and compound your baseline to repeat the cycle whilst sticking to the *plan*. If you progressed to the semifinal at a Grand Slam or any other event, it comes down to the 8 Keys and *your* rate of frequency – be sure to *work it* explicitly and that's the penultimate *key* you're after. But for those of you who fell at the initial hurdle in the 1st Rounds, all is not lost – every player starts somewhere. Let this be a baseline you build upon until that 2nd Round signals that next breakthrough and in time, that Final 8 won't seem as far away as it once was – trust me, I *know* that *feeling* and I *know* what it is like to be in *your* shoes. That's why *your* Black Book and its *Journal* was earnestly made, for you.

The Next Generation Awaits

You have either made it to this point with a full **Black Book** bursting at the seams or you wanted to make your way through to this point *before* you got to *work*. For those of you who are yet to start – it is time to get to *work*. For those of you who are struggling to keep *your* Black Book in one piece that's why the *Journal* is ready when you are. If you chose to use the *Journal* in conjunction with *your* Black Book and both are miraculously in one piece *or* the spines are still holding them intact – send me a picture, *please*!

My [Tennis] Black Book marks a significant milestone in *your* development and in the *Collection*. You can now *navigate* towards the top of the game and whilst *your* Black Book is only the beginning of what I am sure is still more to come, it serves the purpose in preparing you with the necessary building blocks – those that are fundamentally mandated, to **achieve** a Top 10 ranking. Be patient, however, as this won't happen overnight. Inside *your* Black Book and dispersed across the *Collection* are the means to feed your hunger to know *how* and to quench your thirst for the *questions* of *why*. This addition to the *Collection* adds yet another layer to this body of work but in doing so, provides you with a means to *navigate* the *Collection* itself.

What comes next is your next *stage* and/or Level – depending on your rate of progress. For some, you may choose to repeat the same *stage*. For others, you may have opted to complete all *stages* and Levels across the span of the season and next season focus explicitly on the *one* Level. Alternatively, if you *have* broken into the Top 20 on the WTA or ATP Tours, cementing the *last* of the 7 Keys into your game becomes paramount to crossing the threshold whilst for those of you inside the Top 10 with the ambition to *stay* – that's where all 8 Keys come into play. From here, it is relatively simple – as far as the data is concerned. To achieve a Level of *maintenance* and rate of success that **Federer, Nadal, Serena** and **Djokovic** *all* achieved at the heights of their success, this is where the 8 Keys are brought to life – literally, and *why* **Djokovic** remains inside the 8%.

The bedrock of the *Collection* and the 8 Keys is built off the *four* best players of this generation with astounding accomplishments and named in order of their Grand Slam (*singles*) totals, in ascending order, courtesy of all *four* players coming out on top in their own respective categories. But to establish this profound Level of *maintenance* all *four* of these players have heavily contributed to the *Collection* over the years and the underlying *how* to *navigate* and *plan* towards a Top 10 ranking. Each one of these *four* extraordinary players were tracked from the very beginning. More than **ten years** ago when **Djokovic** trailed **Federer** and **Nadal**, the data *knew* this would change. And that's remarkable.

At a similar rate, the data also *knew* why **Serena** was so far ahead of her cohort and why these players were not notching up Grand Slams at the same *rate*. What's fundamental to understand is that 'that' generation has now ended. Whilst the 2024 season officially waved goodbye to **Nadal**, among others, **Serena** and **Federer** said their respective farewells in 2022; **Djokovic** is the *one* left standing. The rate of frequency of **Djokovic's** success, however, has started to slow. A similar pattern occurred with **Federer**, **Serena** and later **Nadal** – their upkeep of the 8 Keys was no longer as robust as it once was. And this is the same challenge **Djokovic** now faces. What once allowed all *four* of these players to dominate eventually was *and* will be responsible for their performance regressions. This rate of maintenance, and its *less* sturdiness this very season, was a clear sign that **Djokovic** is no longer in *'his'* generation but has entered the *next*.

Of course, the next generation has been here for some time but not until **Serena** and **Federer** exited their respective Tours, before **Nadal**, did the *new* contenders take their place alongside **Djokovic**. Having now crossed into the *new* generation, **Djokovic** is no longer *the* most dominant player, however, he still remains in contention. On the WTA Tour, Swiatek and Sabalenka lead the *new* charge whilst Gauff trails and the like of Zheng and Navarro are other 'younger' players inside the Top 10 making moves. Of course, there are several players in the mix, as discussed in the Final 8, and if this season is anything to go by, not *one* player has showed signs of following in **Serena's** footsteps – perhaps *two* but the competition is rife. On the ATP Tour, Sinner and Alcaraz lead whilst **Djokovic** remains within reach of *another* Grand Slam whilst De Minaur was the only 'new' Top 10 arrival of the season.

What's interesting about this *next* generation is so far it comes in *two* – Swiatek *and* Sabalenka, Sinner *and* Alcaraz, with the question remaining which of these *two* respective players on either Tour will move further ahead, or *if* both sets of players will continue to dominate. Irrespective of *how* you view the comparisons and whilst Swiatek and Alcaraz lead their respective Tours Grand Slam haul, both Sabalenka and Sinner had a memorable 2024 season. What is clear, however, is that no *one* player can be attributed to a Level of dominance akin to that of **Federer, Nadal, Serena** or **Djokovic** – although, Swiatek leads on clay with *four* French Open *wins* in *five* years, and Alcaraz leads on grass with *two* Wimbledon *wins* in *two* years.

When **Federer** started his Grand Slam collection, he had a record of *fourteen* Grand Slams to surpass – set by Pete Sampras in the generation *prior*. In contrast, when **Serena** started notching up her Grand Slam *wins,* she had *twenty-two* to surpass – set by Steffi Graf in the generation *prior*. Whilst *all* these players were on Tour at the *same* time, both Sampras and Graf were at the later ends of their careers as the **next generation** was coming through – a similar phenomenon as to *now*. Prior to this record, the singles *all-time* Grand Slam record that is yet to be surpassed (at the time of writing) is that of Margaret Court with *twenty-four* Grand Slams which Graf fell *two* shy of before **Serena's** *one*. Insofar as Sampras' record is concerned, **Federer** was the *first* to surpass this, then it was **Nadal** who would go on to break his rivals record before **Djokovic** would later surpass **Nadal's** Grand Slam total and *lead* the number of *all-time* Grand Slams of his generation – *equal* to that of Margaret Court. This means for the *first* time in history, both the WTA and ATP Tours have an *equal* tally to break which happens to be the *most* Grand Slams *ever* won by a single player in *history* – set by Margaret Court and **Djokovic**, respectively. And that is the almighty task of this **next generation** – can this generation surpass the record set by Margaret Court, or go *one* better than **Serena**, or can this generation match that of the *former* generation and take the game to new heights alike **Federer, Nadal** and **Djokovic** or will the records set by this now *former* generation stand the test of time for another generations, until it is **you** who's ready to make your move? But each generation is not as clear cut as it may seem. As with the respective Grand Slam records overtaken, or fallen short by *one* or *two,* there's an *overlap* between generations and the transition that happens from *one* generation to the **next generation** is underway *now*. This means your time is yet to come *but* the player to surpass **Serena** and the player to come within reach of **Djokovic's** record is putting in the *work* now.

The good news is that there is no definitive rule as to whether the players who are currently leading the **next generation** will be the ones who edge closer to, or even surpass the *equal* Grand Slam record. But *now* is the time for those who aspire to do so, to put in the work. Similarly, there is no firm rule that says the *'next'* record breaker must *win* their *first* Grand Slam by a certain point in time. That is *not* how records are broken. The Level of *maintenance* of all 8 Keys, scientifically established through the *performances* and *results* of *four* of the best players of the 'former' generation – **Federer, Nadal, Serena** and **Djokovic**, *plus* hundreds more, stemmed from the prior generation before *new* parameters could be identified and formed. And now the **next generation** has continued to *add* to the significant data that amplify the power of the 8 Keys to give you the best chance of *success* ever recorded before. The same *identifiers* behind the 8 Keys have continue to align with the current *leading* players on both Tours, be it Swiatek, Sabalenka, Alcaraz and/or Sinner, but with a core differentiator – greater levels of fluctuation. The same Level of maintenance and surefire application held by the likes of **Federer**, **Nadal**, **Serena** and **Djokovic** is yet to be met – the dominance is not as significant, but do not be fooled, all *four* of these players are *gaining*. Whether these players can uphold the *key* rulers is a story that will unfold.

Nothing is absolute – *not* yet. Has the race begun? *Just*. Is it within reach? **Yes**. Now is *your* window. This is *your* window of opportunity. Every generation has their PoI and every generation has that *one* player or *three* that surges ahead courtesy of their knowledge bank *retaining* and *reiterating* the power of the 8 Keys. **Federer** didn't happen overnight. **Nadal** came knocking a few years *later*. **Serena** edged those at their peak close to her PoI. And then came **Djokovic** – there was no time clock. When the 8 Keys are ready, you'll *know*. It *will* happen. But the next generation is now. And this is the *first* generation with access to the 8 Keys – made known for the *first* time in history courtesy of not *one* but *four* of the **greatest players of *all* time**.

But for every *four* there are countless others who have *not* reached the pinnacle of play – but *not* without trying. The data dispersed across the *Collection* surpasses **150,000** inferences – a staggering dataset. Whilst **Federer, Nadal, Serena** and **Djokovic** have collectively led the charge, there are countless others who have held rankings inside the Top 100 on the WTA or ATP Tours but have *not* progressed to the Top 10 or gone on to *win* Grand Slams, but there are also those who have *won* a Grand Slam or a *few* – just not as many, or held a Top 10 ranking. Each of these players over the past decade has *also* contributed to this work. What they *did* and then did *not* do – *how* their Level of maintenance dropped and/or *how* their patterns of play altered and/or *how* their rate of frequency regressed. This is the biggest analysis of its kind to **ever be conducted** on players on the WTA and ATP Tours. This is the most significant investigation to ever have been done on explicitly *how* to progress towards a Top 10 ranking. And *your* Black Book, between your hands, holds the answers. **My [Tennis] Black Book** is an answer in itself – *navigating* the *Collection* to *planning* through *your* Black Book and the adjoining *Journal*, the epitome of *how* to follow in the footsteps of the likes of **Federer, Nadal, Serena** and **Djokovic** has arrived.

Now you can see why *your* Black Book was termed just that – a true **Black Book** in every sense of the meaning *if* your aspiration is to ascend towards the top of the game and be in with a fighting chance to *win* Grand Slams and break a record, or a *few*. Whilst there is no definitive timeline that welcomes in the **new generation**, there are signs – there are markers and then there are windows of opportunity. The 2024 season is one of them. If you're currently on Tour, then this is your time to make your move – the window of opportunity is here. If you're not on Tour and still have several years *before* you complete your *first* decade of play – you now know the signs and that *your* window will also come. Most excitingly, you are the *first* of this **next generation**, or the one to follow, that has been able to access the *first* **Volume** of its kind – **My [Tennis] Black Book** was *made* for you.

Your Black Book is now witness to an entire season that you have completed in conjunction with the *Journal*. *Your* knowledge bank has been set alight with the 8 Keys and the ability to now *navigate* your way through the *Collection*. *Your* questions have been answered whether directly or indirectly through the *planning* and completion of *your* Black Book – prompting you to *ask* questions or use your **superpowers** to progress to that *next* baseline and that *next* peak performance. And I have walked in your shoes, *knowingly*, to meticulously build the penultimate **sidekick** you can take with you and *know* that I am here with you, each step of the way. From this season to the next, *your* Black Book is here to stay. Golden or otherwise, your dreams are worthy of being brought to life and you now have an entire *library* behind you to *navigate* towards a Top 10 ranking. This means it is time to *plan* for next season, to put in the *work* and know that I am in *your* corner. Between each page of *your* Black Book, I remain your willing **sidekick** and when you're ready, when the time has arrived, I'll be waiting. Remember, the **next generation awaits** those courageous enough to dream what's *possible* and hold it in their eyes. Only then does time beckon and I'll be there to welcome you to the next generation – *your* time to play.

What Now?

Visit https://AM8International.com for all of our resources, including the entirety of the *Collection* – your own library to nudge closer towards the Top 10 and to follow the Pathway and The Long Game. Visit Beyond Top 10 Tennis on Spotify Podcasts, Apple Podcasts, YouTube Podcasts, Amazon Podcasts to iHeartRadio. Visit https://Topicthread.com to interact with like-minded people, https://Sprooke.com for random polls or https://PinkOctopus Books.com to learn more about Dr Berge's fictional releases.

Want to get in touch? You can find Dr. B on Topicthread @Dr.B or head to https://BeyondTop10Tennis.com to locate the social links to various accounts, or Subscribe to Dr B's Pack on AM8 International for exclusive guidance towards the Top 10 and/or to *maintain* your hold inside the 8%.

Books by Dr. B

NON-FICTION RELEASES

The Secrets to Optimal Performance Success
(March, 2016)

The Secrets to Optimal Wellbeing
(September, 2016)

The Science of Elite Performance
(March, 2017)

The Secrets to Optimal Coaching Success
(March, 2018)

The Elite Research Method
(October, 2018)

What is Your Game Missing?
(March, 2019)

What is Your Game Missing, Now?
(October, 2019)

What is Your Game Missing, to Win?
(July, 2020)

I am Your Tennis Coaching Guru
(December, 2020)

The 7 Keys to Optimise Your Life
(December, 2021)

My [Tennis] Black Book: Federer, Nadal, Serena, Djokovic, __________

How to Develop a Top 10 Tennis Ranking: The Power of the 8th Key
(December, 2023)

My [TENNIS] Black Book: Federer, Nadal, Serena, Djokovic
Navigate & Plan towards a Top 10 Ranking Vol. 1
(December, 2024)

<u>FICTIONAL RELEASE</u>

The Power of Another Woman
(November, 2022)

Acknowledgements

With each Book I am lucky enough to write and for it to make its way into your hands, it is a milestone itself and if you asked how I have now amassed *thirteen* in the space of *nine* years, I simply do not have the answer other than *you*. Each one of you have kept me going to share more of what it really takes to become the best in the world. And in doing so, there's an entire team behind the scenes that allows me to keep going – *thank you* endlessly. From each one of you who has tuned in to Beyond Top 10 Tennis – *thank you*. For those who have visited and keep coming back to AM8 International – *thank you*. For those of you who have already embarked on the *Collection – thank you*. For those who continue to share what AM8 International has to offer and its relentless pursuit – *thank you* immeasurably. And yet a *thank you* doesn't seem enough because you allow me to keep writing.

***

Malibu & Maldive, my sidekicks throughout. Being your Mum is my greatest privilege.

Bibliography

The *Collection*

Berge, A.M. (2023). How to Develop a Top 10 Tennis Ranking: the Power of the 8th Key. Sydney, Australia: AM8 International.

Berge, A.M. (2021). The 7 Keys to Optimise Your Life: using tennis to develop behaviours that deliver optimal performances in play and in life. Sydney, Australia: AM8 International.

Berge, A.M. (2020a). What is Your Game Missing, to Win? How tennis forms the blueprint to remove youth and elite injury with new identifiers to solve Replicated Success for the best tennis players in the world. Columbia, SC: AM8 International.

Berge, A.M. (2020b). I am Your Tennis Coaching Guru: a 10 year pathway of lessons and planners for junior to amateur players towards a top 10 ranking. Middletown, DE: AM8 International.

Berge, A.M. (2019a). What is Your Game Missing? The scientific truth in breaking down tennis' elite. Middletown, DE: AM8 International.

Berge, A.M. (2019b). What is Your Game Missing, Now? The elite tennis player advantage. Middletown, DE: AM8 International.

Berge, A.M. (2018a). The Secrets to Optimal Coaching Success: the role of experience, optimal performance practices and outcomes in the real-world. Columbia, SC: AM8 International.

Berge, A.M. (2018b). The Elite Research Method: a methodological framework applied in the real-world. Columbia, SC: AM8 International.

Berge, A.M. (2017). The Science of Elite Performance – The World Awaits: an infusion of science, education and communication: a complete learning structure on how elite coaches and athletes become the best in the world. Lexington, KY: AM8 International.

Berge, A.M. (2016a). The Secrets to Optimal Performance Success: a comprehensive discussion on developing elite coaches and players. Columbia, SC: AM8 International.

Berge, A.M. (2016b). The Secrets to Optimal Wellbeing: a 12 week frecstyle nutrition and functional movement fitness guide. Columbia, SC: AM8 International.

FROM THE WORLD'S LEADING SCIENTIST BEHIND
'HOW TO DEVELOP A TOP 10 TENNIS RANKING'

MY [TENNIS]

FEDERER. NADAL.

BLACK BOOK

Vol. I

SERENA. DJOKOVIC.

NAVIGATE & PLAN TOWARDS A TOP 10 RANKING

DR. ASHLEY M. BERGE, PHD

AM8 International

Beyond Top 10 Tennis

Beyond Top 10 Tennis Rankings

Remember to reach for the Stars,
Golden or otherwise, persevere,
You can & You are capable.
Go for it!

At the ready,
Dr. Berge.

AM8International.com